Representing Direction
in Language and Space

EXPLORATIONS IN LANGUAGE AND SPACE

Series editor
Emile van der Zee, *University of Lincoln*

PUBLISHED

Representing Direction in Language and Space
Edited by Emile van der Zee and Jon Slack

IN PREPARATION

Functional Features in Language and Space:
Insights from Perception, Categorization, and Development
Edited by Laura Carlson and Emile van der Zee

Representing Direction in Language and Space

Edited by

EMILE VAN DER ZEE
and JON SLACK

UNIVERSITY PRESS

This book has been printed digitally and produced in a standard specification in order to ensure its continuing availability

OXFORD
UNIVERSITY PRESS

Great Clarendon Street, Oxford OX2 6DP
Oxford University Press is a department of the University of Oxford.
It furthers the University's objective of excellence in research, scholarship,
and education by publishing worldwide in
Oxford New York
Auckland Cape Town Dar es Salaam Hong Kong Karachi
Kuala Lumpur Madrid Melbourne Mexico City Nairobi
New Delhi Shanghai Taipei Toronto
With offices in
Argentina Austria Brazil Chile Czech Republic France Greece
Guatemala Hungary Italy Japan South Korea Poland Portugal
Singapore Switzerland Thailand Turkey Ukraine Vietnam

Oxford is a registered trade mark of Oxford University Press
in the UK and in certain other countries

Published in the United States
by Oxford University Press Inc., New York

© Editorial matter and organization van der Zee and Slack
© Chapters their several authors

The moral rights of the author have been asserted

Database right Oxford University Press (maker)

Reprinted 2008

All rights reserved. No part of this publication may be reproduced,
stored in a retrieval system, or transmitted, in any form or by any means,
without the prior permission in writing of Oxford University Press,
or as expressly permitted by law, or under terms agreed with the appropriate
reprographics rights organization. Enquiries concerning reproduction
outside the scope of the above should be sent to the Rights Department,
Oxford University Press, at the address above

You must not circulate this book in any other binding or cover
And you must impose this same condition on any acquirer

ISBN 978-0-19-926019-5

Contents

Preface	vii
List of Figures	viii
List of Tables	xiii
Abbreviations and Keywords	xiv
Contributors	xv

1. The Representation of Direction in Language and Space 1
   Jon Slack and Emile van der Zee
2. Axes and Direction in Spatial Language and Spatial Cognition 18
   Barbara Landau
3. Vectors across Spatial Domains: From Place to Size, Orientation, Shape, and Parts 39
   Joost Zwarts
4. Vector Grammar, Places, and the Functional Role of the Spatial Prepositions in English 69
   John O'Keefe
5. The Unique Vector Constraint: The Impact of Direction Changes on the Linguistic Segmentation of Motion Events 86
   Jürgen Bohnemeyer
6. Defining Spatial Relations: Reconciling Axis and Vector Representations 111
   Laura Carlson, Terry Regier, and Eric Covey
7. Places: Points, Planes, Paths, and Portions 132
   Barbara Tversky
8. Ontological Problems for the Semantics of Spatial Expressions in Natural Language 144
   Pierre Gambarotto and Philippe Muller
9. Change of Orientation 166
   Hedda Schmidtke, Ladina Tschander, Carola Eschenbach, and Christopher Habel
10. How Finnish Postpositions See the Axis System 191
    Urpo Nikanne
11. Directions from Shape: How Spatial Features Determine Reference Axis Categorization 209
    Emile van der Zee and Rik Eshuis

12. Memory for Locations Relative to Objects: Axes and the
    Categorization of Regions   226
    Rik Eshuis
13. Spatial Prepositions, Spatial Templates, and 'Semantic' versus
    'Pragmatic' Visual Representations   255
    Kenny Coventry

References   268
Index   281

Preface

The domain of language and space is an important interstitial area, where disciplines from linguistics, psychology, computational approaches to cognition, and the brain sciences fruitfully interact. For this reason, it has proved to be one of the most dynamic areas of cognitive science. This is the first book in a series which aims to capture this interdisciplinary ferment.

Each book in the series begins its life in a workshop on an important aspect of the subject. The workshops provide the opportunity for scholars in different disciplines to discuss their approaches, findings, and methods. Both they and the books are intended to provide a forum for the exchange of ideas and recent developments, as well as for debates on points at issue. Those papers selected for publication are revised in the light of discussion, peer review, and reports by the publisher's external referees. The present volume had its beginnings in a meeting held at the University of Lincoln in July 2000. A similar meeting on functional features held at the University of Notre Dame in 2001 will form the basis of the next book in the series. A workshop on granularity held in July 2002 at the University of Bielefeld will provide the inspiration for the third.

Many thanks to all those who have made the series possible, especially the participants and the authors. I am grateful to Jon Slack for co-organizing the first workshop and co-editing this book, and to the staff at OUP for bringing the work to its final printed form.

Emile van der Zee

Lincoln

List of Figures

1.1	The mappings or interfaces that are studied in the area of Language and Space	2
1.2	The mappings or interfaces that are the focus of this book	2
1.3	Direction defined in relation to regions, in combination with basic geometric axioms	6
1.4	Direction defined on the basis of lines, points, and half-lines, in combination with basic geometric axioms	6
1.5	A vector has a length (encoding distance), and an orientation (encoding direction)	7
1.6	An axial system allows for the directional specification of a point in terms of categories and in terms of a more refined encoding of space	8
1.7	By relating a vector to an intrinsic axis a vector's head can encode direction	12
2.1	Schema of results for two spatial tasks, the Memory task and the Language task	25
2.2	Samples of drawings by a child with Williams syndrome (WS) and a normally developing child matched for mental age	28
2.3	Mean percent correct for locations that are distant from the Reference object along vertical and horizontal axes	29
2.4	Locations for which children produced vertical positive lines (e.g. *above, over, on, off*, etc.)	30
2.5	Children's dot placement locations when asked to put a dot *above/over/on top of* the black square	32
3.1	*The plow behind the tractor*	42
3.2	*Near the tractor, next to the tractor, between the tractor and the barn*	43
3.3	Head–tail path	44
3.4	*Around the tree*	45
3.5	*Over the house*	45
3.6	*Towards the city*	45
3.7	Place vector	47
3.8	Axis vector	47
3.9	Projection of a vector on the upward direction	49
3.10	A stick and one of its ends	53
3.11	(a) *X in front of*; (b) *on the front of*; (c) *in the front of y*	55

3.12	A circular path	55
3.13	*X circles y*	56
3.14	*X rotates*	56
3.15	*A round disk*	57
3.16	*A round ring*	57
3.17	*Spiral staircase*	58
3.18	Centered path of a river	59
3.19	Head–tail path of a river	59
3.20	The coast	63
3.21	The Chinese wall	63
3.22	A zigzag path	64
3.23	Orientation *straight*	65
3.24	Shape *straight*	65
3.25	'Growing askew'	65
3.26	'Growing curved'	66
3.27	To bend to the left	67
4.1	Three examples of the fields of boundary vector cells	75
4.2	The place identified by the preposition *beyond*	77
4.3	The place identified by the preposition *below*	77
4.4	(a) Acceptability ratings for locations *below the O* (b) Acceptability ratings for the preposition *above*	78
4.5	(a) The place identified by the preposition *behind* (b) The semantic field of *farther behind*	79
4.6	The place identified by the preposition *under*	80
4.7	*Under* the horizontal shelf	80
4.8	The place identified by the preposition *by*	81
4.9	The semantic field of *between* the square and the circle	83
5.1	First and last frame of ECOM B5	88
5.2	First and last frame of ECOM C6	88
5.3	Motion vectors in ECOM C6	91
5.4	The coding of direction in motion event descriptions	97
5.5	The relationship between the AUC and the UVC	99
5.6	The representation of direction in different frames of reference	100
5.7	Implicated path curvature: *The ant crawled up over the table*	106
5.8	Path shape underdetermines direction	109
6.1	'The coin is above the piggy bank'	112
6.2	'Above' spatial template	114
6.3	Illustration of offsetting the functional part from its center-of-mass	118
6.4	The attentional vector-sum (AVS) model	120

6.5	Alignment of the proximal and the center-of-mass orientations	122
6.6	Critical placements, ratings, and model predictions around a rectangular reference object	123
6.7	Placements and mean acceptability ratings around the tall triangle reference object	125
6.8	Empirical data and model predictions for critical placements above a rectangular reference object	128
8.1	Mereo-topological relations	152
8.2	Points constructed from regions, and an example of an induced direction	153
8.3	Relations between spheres	155
8.4	A spatio-temporal interpretation of overlap	158
8.5	Illustration of temporal relations	159
8.6	Motion verbs	159
9.1	Examples of spatial change	168
9.2	Half-lines induced by the line of sight	171
9.3	(a) A half-line that is part of another half-line (b) Parallel half-lines	173
9.4	Configurations of two half-lines that are part of the same straight line	174
9.5	P and Q lie on the same side of g	174
9.6	(a) same orientation; (b) opposite orientation; (c) different orientation	176
9.7	Equidirected half-lines	177
9.8	Depictions of cases of *abbiegen* ('turn off')	184
9.9	(a) An example of *nach rechts abbiegen* ('turn to the right') (b) The corresponding direction diagram	185
9.10	Example for *schräg nach rechts abbiegen* with a corresponding direction diagram	187
9.11	An illustration of dbtw(r_1, r, r_2) with a corresponding direction diagram	188
9.12	Correspondence between dbtw(r_1, r_2, r_3) and start (r_4) $<_{r_4} P$ and $P <_{r_4} Q$	189
10.1	Theme, PLACE, and reference object. Illustration of the sentence *John is standing next to the table*	193
10.2	Two rockets in motion	201
10.3	Flying helicopters	202
10.4	'The fly is flying beside the bee'	202
10.5	(a) 'The bee is flying in front of the fly and the beetle (is flying) beside it'	

10.5	(b) Buick ajoi Volvon eteen/edelle Buick drove Volvo+GEN in-front-of+ILL/in-front-of+ALL (c) Buick ajoi Volvon edellä/edessä Buick drove Volvo+GEN in-front-of+ADE/in-front-of+INE	203
10.6	(a) Axial system: one vertical and two horizontal axes (b) Buick ajoi Volvon taakse/perään Buick drove Volvo+GEN behind+TRA/in-front-of+ILL	204
	(c) Buick ajoj Volvon takana/perässä Buick drove Volvo+GEN behind+ESS/behind+INE	205
11.1	The reference objects used in experiments 1 and 2	211
11.2	It is assumed here that the distribution of the intrinsic regions and object sides that correspond to Dutch directional prepositions and nouns is based on the position and orientation of the main axis and two orthogonal axes of the closest fitting cuboid or bounding box around a reference object	212
11.3	Cuboid axis categorizations for intrinsic reference to regions around reference objects in the horizontal plane based on the use of Dutch directional prepositions	214
11.4	Predictions of sticker placements corresponding to Dutch directional nouns based on the intrinsic axes of the reference objects	217
11.5	Predictions of sticker placements corresponding to Dutch directional nouns assuming the placements to be as close as possible to the side of a closest fitting cuboid	218
11.6	The stimuli that were used in experiment 3	222
12.1	One of the stimulus configurations used by Hayward and Tarr (1995)	229
12.2	Encoding and reproduction of an intrinsic spatial relation	234
12.3	The two Referents used in experiment 1	240
12.4	Illustration of the treatment of Figure location estimates	242
12.5	Reaction times for both objects and each direction	243
12.6	Standard ellipses for each location relative to the Referents	245
12.7	The Referent used in experiment 2	247
12.8	Standard ellipses for each location around the Referent	249
12.9	Construction of the regional error measure in experiment 2	250
13.1	Acceptability ratings for the spatial relation *above*, according to Logan and Sadler (1996) as modified by Carlson-Radvansky and Logan (1997)	257

13.2 Contrary to predictions by spatial template theory acceptability ratings for *over* and *above* are related to the degree of rotation and the functionality of the Figure   260
13.3 Figure and Ground object relations aligned in the same way in different reference frames, or not aligned in different reference frames   261
13.4 Interaction between geometry, function, and preposition set found in Coventry, Prat-Sala, and Richards (2001) when frames of reference coincide   262
13.5 Interaction between geometry and preposition set found in Coventry, Prat-Sala, and Richards (2001) when frames of reference conflict   264
13.6 Interaction between function and preposition set found in Coventry, Prat-Sala, and Richards (2001) when frames of reference conflict   264

List of Tables

5.1	Direction terms according to frame of reference	100
9.1	Verbs of motion and verbs of orientation combined with spatial adverbial phrases	180
9.2	Verbs of change of orientation combined with spatial adverbial phrases	183
10.1	Features of spatial relations by Jackendoff and Landau	192
10.2	The case paradigm of the postpositions *ete-* and *taka*	196
10.3	Finnish postpositions that indicate a spatial PLACE	199
11.1	Correspondence between directional axis marking and lexical concepts describing reference axis categorization for intrinsic directional reference by Dutch native speakers	221
12.1	Error types for each Referent	244
12.2	Regional errors in experiment 2	251

Abbreviations and Keywords

Abbreviations

ADE	adessive case	NP	noun phrase
ALL	allative case	P	preposition
AUC	Argument Uniqueness Constraint	PAR	partitive case
		PC	proximal center of mass
AVS	attentional vector sum	PP	prepositional phrase
BB	bounding box	RBC	recognition by components
BVC	Boundary Vector Cell	REFL	reflexive
Dir	directionality	SFC	spatial feature categorization
ECOM	Event Complexity	SG	singular
ESS	essive case	TOP	topic
FoR	Frame of Reference	TRA	translative case
GB	Government and Binding	UVC	Unique Vector Constraint
GEN	genitive case	V	verb
ILL	illative case	VP	verb phrase
INE	inessive case	WS	Williams syndrome
LFG	Lexical Functional Grammar		

Keywords

Words that play an important role in the theme of the book or a particular book chapter are represented in **bold** when mentioned for the first time, or when formally or informally defined. References to concepts are in small capitals, so that e.g. the concept for 'apple' is printed as APPLE.

Contributors

JÜRGEN BOHNEMEYER is Assistant Professor of Linguistics at the University of Buffalo, United States. His research focuses on the representation of complex events in language and cognition.

LAURA CARLSON is Professor of Psychology at the University of Notre Dame, United States. Her research interests focus on the interface between language and perception in the domain of spatial relations.

KENNY COVENTRY is Principal Lecturer at the Centre for Thinking and Language, University of Plymouth, United Kingdom. His research topics are spatial language and spatial representation, and spatial prepositions in particular.

ERIC COVEY is a Ph.D. student in the Psychology Department at the University of Notre Dame, United States. His research interests include the representation and processing of spatial information.

CAROLA ESCHENBACH is Senior Research Scientist and lecturer at the Department of Informatics at the University of Hamburg, Germany. She investigates how the structure of the world is reflected in natural language semantics.

RIK ESHUIS is a Ph.D. student in the Doctoral Program in Cognitive Science at the University of Hamburg, Germany. His research topic is memory for object location, and its relation to spatial language.

PIERRE GAMBAROTTO is a Ph.D. student in Artificial Intelligence at the Computer Science Laboratory of the Université Paul Sabatier of Toulouse, France. His research focuses on the formal representation of spatial information.

CHRISTOPHER HABEL is Professor of Computer Science and Linguistics at the University of Hamburg, Germany. His research topics include the interaction of semantics and spatial cognition from a cognitive as well as from a mathematical point of view.

BARBARA LANDAU is Professor of Cognitive Science at Johns Hopkins University, United States. Her research concerns the representation and acquisition of spatial cognition and spatial language, and the interactions of these two systems of knowledge in the mind and brain.

PHILIPPE MULLER is Maître de Conferences in Computer Science at the Université Paul Sabatier of Toulouse, France. His research interests include natural language semantics, the processing of discourse and dialog, and spatial reasoning.

URPO NIKANNE is Professor of Finnish at the Åbo Akademi University in Turku, Finland. His research focuses on Finnish grammar and the theory of conceptual semantics.

JOHN O'KEEFE is Professor of Cognitive Neuroscience at the Department of Anatomy and Developmental Biology and the Institute of Cognitive Neuroscience, University College, London, United Kingdom. His research focuses on the neuronal basis of spatial memory and navigation.

TERRY REGIER is an Associate Professor of Psychology at the University of Chicago, United States. His research concerns the psychological bases of linguistic meaning, with a particular focus on the domain of spatial relations.

HEDDA RAHEL SCHMIDTKE is a Research Assistant and doctoral student at the Department of Informatics at the University of Hamburg, Germany. Her research interests focus on the ontology of space.

JON SLACK heads the Department of Psychology at the University of Lincoln, United Kingdom. His main research interests are in neural network modeling and representational theory.

LADINA TSCHANDER is a Research Assistant and doctoral student at the Department of Informatics at the University of Hamburg, Germany. She currently prepares a Ph.D. thesis in the field of lexical semantics.

BARBARA TVERSKY is Professor in Psychology at Stanford University, United States. Her research interests include picture memory and pictorial representations, imagery, spatial thinking, spatial language, cognitive maps and graphs, recollections and eyewitness testimony, systematic distortions in memory, HCI, and mental models constructed from text.

EMILE VAN DER ZEE is Senior Lecturer in Psychology at the University of Lincoln, United Kingdom. His research topics include the representation of space for language, and the representation of (lexical) conceptual structure for talking about space.

JOOST ZWARTS is a post-doctoral researcher at the Utrecht Institute of Linguistics OTS of Utrecht University, The Netherlands. His research interests are in the domain of syntax and semantics, and in the grammar of Endo, a Nilotic language that he has studied for some years when working in Kenya with SIL.

1

The Representation of Direction in Language and Space

JON SLACK and EMILE VAN DER ZEE

This book focuses on how direction is represented in language, and how direction is represented in spatial terms for the purpose of language. Two issues are addressed in particular. The first issue concerns the constituents of language encoding direction. Which linguistic constituents encode direction, and how do they do it? The second issue concerns the possible vehicle for representing a direction such as *in front of the car*. What are the spatial distinctions underlying such linguistic expressions? In this chapter we address these issues, and discuss each of the book's chapters in relation to them.

Why is direction representation in language an interesting issue in Cognitive Science? In order to answer this question we need to start by considering the relation between language and spatial representation. Every language contains fixed expressions (words and idioms) that refer to spatio-temporal configurations such as *around, yesterday,* and *from x to y* (where x and y can be spatial variables like *Helsinki* and *Amsterdam*, or temporal variables such as *23 June* and *26 June*). The research area of **Language and Space**[1] studies the properties of the mapping or the interface between space-time and the subset of linguistic expressions referring to spatio-temporal configurations (see e.g. Bloom *et al.*, 1996; Coventry and Garrod, forthcoming; Freksa, Habel, and Wender, 1998; Herskovits, 1986; Miller and Johnson-Laird, 1976; Olivier and Gapp, 1998; Talmy, 2000; van der Zee and Nikanne, 2000). More specifically, the research area of Language and Space focuses on the way space-time maps to the formal or cognitive representation of space-time, and the way in which such space-time representations map to spatial language or $L_{spatial}$. $L_{spatial}$ is the subset of linguistic expressions describing such aspects as object location, object configuration, and object movement. Figure 1.1 depicts the mappings involved (see also Jackendoff, 1996b; Peterson *et al.* 1996).

---

[1] Throughout this book key concepts are represented in a bold font. This occurs only at those locations in each chapter at which a key concept is introduced or defined.

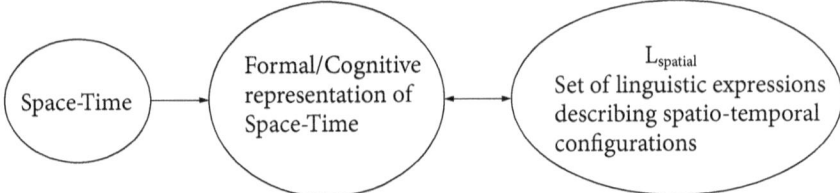

FIG. 1.1 The mappings or interfaces that are studied in the area of Language and Space.

Within this general framework, this book focuses mainly on a subset of the mappings in Fig. 1.1: the way direction as a derived property of space-time relates to the formal or cognitive representation of direction, and the way in which the latter maps onto $L_{directional}$, or the subset of linguistic expressions referring to direction. These more specific mappings are portrayed in Fig. 1.2.

Let us look at Fig. 1.2 in more detail. The first thing to note is that in principle the mappings are **many-to-many mappings**. A given space-time element can map onto many possible formal or cognitive representations of direction, and vice versa. In addition, different representations of direction can map onto the same linguistic element, and vice versa. But, are there any **constraints on the many-to-many mappings**? Let us consider in what way the mappings are constrained by looking at the kinds of entities that comprise each of the domains in Fig. 1.2.

The space-time domain comprises all the space-time configurations that are permitted by the laws of physics. The next domain contains all the formal and cognitive representations of direction that are consistent with—respectively— the principles of mathematics and logic, and the properties of the brain. This means that the domain of formal and cognitive representation is constrained by these sets of principles, but, how about the domain of $L_{directional}$? What are the **linguistic constituents that encode direction**? Let us first look at the formal linguistic devices that are available to represent direction.

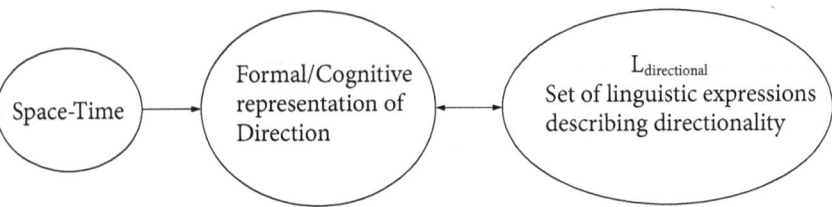

FIG. 1.2 The mappings or interfaces that are the focus of this book.

The following examples show that in English all major syntactic categories can encode direction:

(1) Mary *faced/entered* the train station.
(2) In which *direction* was she oriented/did she go?
(3) The horse walked *north*.
(4) The *overturned* car looked rusty.
(5) The tractor is *behind* the barn.

Example (1) shows that verbs can be used to express the direction in which an object is oriented, or is moving. Example (2) quite literally shows that the noun *direction* encodes direction. Example (3) demonstrates that adverbs—in combination with a motion verb—can encode the direction in which an entity is moving. Example (4) shows that adjectives can encode the direction in which an object is oriented. And example (5) shows that prepositions can describe the location of a Figure (here: *the tractor*) in a particular direction from a Ground (here: *the barn*).

One other possible way in which direction can be formally encoded in language is by case marking:

(6) Der Hund läuft in den Park.
    The dog walks in the+ACCUSATIVE park.
    The dog walks into the park.
(7) Der Hund läuft in dem Park.
    The dog walks in the+DATIVE park.
    The dog walks around in the park.

The German examples in (6) and (7) demonstrate that the simple contrast between accusative case and dative case can encode—respectively—whether the path that the Figure object ('the dog') is traveling along is directed, as in (6), or undirected, as in (7). In fact, different cases can encode directions quite specifically (Svorou, 1994). In Finnish, for example, the ablative and elative cases can be used to encode a direction starting at some source, respectively *from on . . .* and *from at/in . . .*, whereas the allative and illative cases can be used to refer to a direction ending at some goal, respectively *to on . . .* and *to at/in . . .* (Nikanne, 1990). Case can even be used to select specific goals or sources, for example, venitive case in Kuliak in northeastern Uganda can be used to denote movement towards the speaker or the deictic centre (Heine, Claudi, and Hünnemeyer, 1991). What the above examples show is that the formal means available within the design of languages for encoding direction—syntax and morphology—do not impose significant constraints on encoding directionality in language.

Are there any **constraints on linguistic directional encoding** that follow from the structure of language itself? Consider the following examples:

(8) *Sally walked out of the library from the reception to the entrance (Bohnemeyer, Ch. 5).[2]

(9) *Paul geht rechtsherum (Schmidtke, Tschander, Eschenbach, and Habel, Ch. 9).
Paul walks clockwise.

(10) ??Raketti B on/kulkee raketin A yläpuolella/yllä/päällä. (Nikanne, Ch. 10)
rocket B is/go+3SG rocket+GEN A above/over.
Rocket B is/goes above rocket A.

Examples (8) through (10) are taken from some of the chapters in this book. These ill-formed examples demonstrate that there are constraints on linguistic directional encoding. Although there are no syntactic or morphological reasons why these sentences are unacceptable, they are nevertheless unacceptable to the native speakers of the languages these examples are taken from. The theoretical reasons given for why the above examples are unacceptable relate to constraints on the syntax-to-meaning mapping or to constraints on the well-formedness of sentence meaning. Bohnemeyer argues that there is a universal constraint on directional encoding, forbidding a mapping of multiple directions to a single clause, as in example (8) (see Ch. 5 for a more precise version of this idea). In Ch. 9 Schmidtke, Tschander, Eschenbach, and Habel argue that it is not possible in German to combine a verb describing object translation with an adverb referring to a rotation of the same object, as in (9). Finally, Nikanne shows in Ch. 10 that in Finnish, postpositions referring to the vertical dimension cannot encode the location of one object in relation to another object if both objects are moving upwards, which is why (10) is an inappropriate way of describing object location in such a situation. But, whatever the theoretical reasons given, the examples point out that the mapping of directional information from the domain of formal or cognitive representation to $L_{directional}$ is considerably constrained by the way in which linguistic structure can encode direction—both universally and language specifically.

As we go from left to right across the domains in Fig. 1.2, information is lost in the sense that formal and cognitive representations of time-space necessarily underspecify the continuous nature of space-time. Moreover, information

---

[2] Note that this sentence must be read without any pauses, or without any changes of intonation. If a pause is inserted, e.g. between 'library' and 'from', the sentence can be read as a conjunction—thus satisfying the unique vector constraint (Bohnemeyer, Ch. 5).

is 'lost' by formally or cognitively representing space-time, and by linguistically representing direction on the basis of formal or cognitive representations of direction. Such ideas of information loss are referred to in the literature as the schematization of information in language or as the filtering out of information for language, compared to the information available in the space-time domain (for specific ideas on this topic, see Herskovits, 1998; Landau and Jackendoff, 1993; Nikanne, 2000*a*; Talmy, 2000).

There is another aspect of the mappings in Fig. 1.2 that deserves attention. The fact that the arrow between the formal or cognitive representation of direction and $L_{directional}$ is bi-directional indicates that the formal or cognitive representations of direction and $L_{directional}$ mutually constrain each other. This brings us to the second issue addressed in this book: what are the **spatial distinctions underlying linguistic directional representations**? The spatial distinctions that must be present at the level of formal or cognitive representation for $L_{directional}$ to exist in a meaningful way are the spatial primitives encoding directionality, and a reference system able to interpret directionality. This idea is motivated below. But, let us first consider why it is assumed here that the primitives underlying the meaning of $L_{directional}$ are spatial primitives, such as axes, vectors, etc.

There is no a priori reason why the primitives encoding direction must be spatial entities such as axes or vectors. The currency in which direction can be formally or cognitively encoded might as well be algebraic or propositional. The purpose of this book is not to address the ontological status of a formal or cognitive representation of direction (addressing such an issue would be similar to considering the arguments that played a role in the imagery debate; see e.g. Eysenck and Keane (1995) for a summary). We merely wish to observe here that all the authors in this book adopt spatial primitives such as axes and vectors to represent direction. Since direction is a geometrical notion, it seems most convenient to assume that direction is represented in terms of geometrical as opposed to algebraic or propositional distinctions.

What spatial primitives are assumed to encode direction? The authors in this book have different ideas as to what these primitives are: vectors, half-axes, half-lines, or topological distinctions. There are good arguments for assuming each of these entities, even though the representational properties of these entities vary (mainly in terms of other possible distinctions that these entities encode; e.g. vectors not only encode direction, but also distance). Gambarotto and Muller, Ch. 8, provide a cogent argument for starting with regions of space as the primitives for building a geometric representation of direction (see Fig. 1.3). Spatial language relates to objects, and objects occupy regions of space, so it makes sense to begin with regions in building a formal semantics for

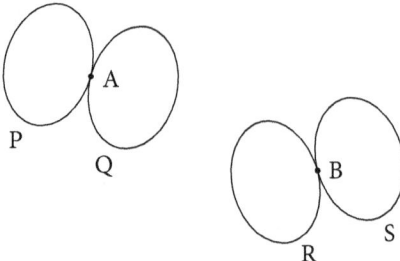

FIG. 1.3 Where regions P and Q touch, a point (A) can be defined, and where regions R and S touch a point (B) can be defined. Paraphrasing Gambaretto and Muller (Ch. 8), it is possible to say that a relation between two points (here: A and B) defines a direction. If A and B form an ordered set of points, orientation is also defined. Here direction is thus eventually defined in relation to regions, in combination with basic geometric axioms.

$L_{directional}$. By defining key **topological relations** on such primitives, Gambarotto and Muller show that their geometric framework can also capture orientation and distance. A key advantage of taking regions of space as spatial primitives is that it is congruent with the notion of receptive fields and much of the research in vision that looks at visuo-spatial representation (Zeki, 1990).

Taking an axiomatic approach Schmidtke *et al.*, Ch. 9, construct a geometric framework from the primitive elements of points, straight lines and **half-lines** (see Fig. 1.4). While keeping the amount of axiomatic structure to a minimum, they define the all-important relation of equidirection that allows them to account for the notions of change of orientation and direction implicit in the

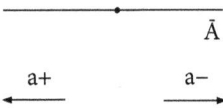

FIG. 1.4 It is possible to divide a line segment (Ā) into to two half-lines at some arbitrary point. Paraphrasing Schmidtke *et al.*, it is possible to say that the two half-lines are converse (i.e. they specify different directions within one dimension, such as *left* and *right*), and that any line parallel to Ā (or a+ and a−) is equidirected with Ā (or a+ and a−), and any line that is not parallel to Ā has a different direction than Ā. By also introducing the notion of orthogonality, Schmidtke *et al.* (Ch. 9) are able to define a system of orthogonal axes, based on the geometry of half-lines. In this way direction is thus defined on the basis of lines, points, and half-lines, in combination with basic geometric axioms.

meanings of a range of German expressions based on verbs of motion. They do not attempt to provide a complete formal basis for $L_{directional}$, but the assumption is that a formal system that can capture expressions of changes in orientation ought to be capable of formalizing the underlying range of orientation phenomena.

The geometric frameworks developed by Gambarotto and Muller, and by Schmidtke *et al.*, are alternative formal representations for direction and do not claim to be cognitive representations. Schmidtke *et al.* explicitly state that their consideration of half-lines as a means for representing direction is not necessarily the solution of the mind/brain for representing direction.

An emphasis on the formal rather than cognitive representation of space is also characteristic of the formal linguistic approach of Zwarts, in Ch. 3. Zwarts develops a formal semantic model of $L_{directional}$ based on **vectors** (see Fig. 1.5). This model integrates the semantics of four spatial domains—size, orientation, shape, and spatial parts. Zwarts consistently and convincingly shows that spatial relationships and their modifications in each of these domains can be explained with a vector-based approach. Vector representations are based on a point geometry in which both direction and distance are regarded as basic. The location of an entity within this type of geometry is given by a vector from a known location. The known location is referred to as the *tail*, or origin, and the location it 'points to' is called the *head*. In order to assign values to the distance and direction components of a vector, a vector representation requires some form of coordinate system (Cartesian, or polar). The import of Zwarts's approach is that it provides a unified semantic basis for a diverse range of spatial domains.

Bohnemeyer (Ch. 5) argues that a vector-based representation is also required to capture the constraints on linguistic motion event coding. As part of a project by the Cognitive Anthropology group at the Max Planck Institute in Nijmegen looking at the linguistic representation of complex events across a diverse range of human languages, Bohnemeyer's work focuses on the linguistic representation of motion events. The key proposal is that there is a one-

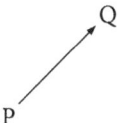

FIG. 1.5 Here P is the tail and Q is the head of vector PQ. Vector PQ has a length (encoding distance), and an orientation (encoding direction).

to-one correspondence between the vector representation of a motion event and the clausal structure of its linguistic specification. This does not mean that the descriptions of the events are always faithful and complete, but rather that a motion event clause is limited in its capacity to express directional information. There is no claim that linguistic descriptions of motion events are generated from underlying vector representations. The vector representation is used here as a descriptive device rather than an explanatory mechanism. Some cognitive scientists may regard the identified constraint as the starting point for seeking just such an explanation and a vector representation might provide the basis for the explanatory framework that emerges. This, however, is not Bohnemeyer's goal. As with Zwarts's use of vectors, Bohnemeyer employs them as formal specifications rather than as constituents of some causal framework. In contrast, Carlson, Regier, and Covey, and also O'Keefe, interpret vectors as cognitive entities (about which more below).

Some authors employ a system of **axes** to account for the semantics of $L_{directional}$. A set of two or more orthogonal axes is assumed to allow for a categorical and if necessary a more refined encoding of space (see Fig. 1.6). Schmidtke *et al.* assume a categorical partitioning of space on the basis of axes based on half-lines. Their axial system can capture direction but not distance. To incorporate distance would require basic metrical concepts to be added to the framework. In this sense Schmidtke *et al.*'s framework is clearly different from a vector-based approach in which distance information is explicitly encoded (see above).

Landau, in Ch. 2, argues quite strongly for axes as representing direction for the purpose of language. On the basis of an extensive literature review Landau argues that there are separate cognitive systems for representing axes and directions. One argument for this idea derives from observations of chil-

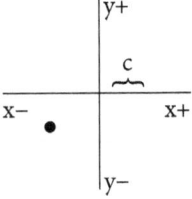

FIG. 1.6 An axial system (axes x and y, or half axes x+, x–, y+, and y–) allows for the directional specification of a point in terms of categories (half-planes, like x+, or quadrants, like x– and y–) and—if required—in terms of a more refined encoding of space (e.g. by defining coordinates ('c') on the axes).

dren with Williams syndrome. Children with Williams syndrome seem to have normal verbal abilities, but impaired spatial abilities, making their cognitive behavior very interesting for those interested in the relation between linguistic and spatial representation. Like other children, Williams syndrome children's memory for object location is good at intrinsic object axes. However, unlike other children, Williams syndrome children's comprehension of linguistically encoded direction (within one axis) is not good, leading to above/below and left/right confusions. This observation is highly important, in that it separates axial representation from directional representation. Landau furthermore gives arguments for why she feels that vector representations do not play a role in directional linguistic reference, but only in spatial tasks.

Nikanne, Ch. 10, uses an axial system to capture the semantics of Finnish postpositions. Finnish allows certain aspects of place meaning to be expressed through the use of prepositions that have no grammatical or lexical correspondences in English. This is because English prepositions refer to fixed spatial locations and do not accommodate motion. In contrast, the Finnish p-position system (prepositions and postpositions) distinguishes between stationary and moving objects, although only in certain dimensions. To capture the semantics of Finnish postpositions, Nikanne argues that it is necessary to employ an axial system that integrates both a number of spatial dimensions and orientation. What he proposes is a three-dimesional axial framework in which the axes are hierarchically organized—the three dimensions can be differentiated as the vertical dimension and the two-dimensional plane, and the latter can be further differentiated as a front–back axis (one-dimensional) and a second axis that is perpendicular to it. In this framework, the vertical axis is special in that its orientation is defined relative to some global directional system. In this sense, the orientation of the vertical axis is given and the other two dimensions are specified relative to it. The global directional system used by Nikanne is an example of a reference frame.

Second to the spatial primitives representing direction, the notion of a **reference frame** is of crucial importance to understanding directional reference in language. A reference frame consists of a set of axes, encoding directional distinctions, such as left–right. In contrast to a set of axes as in Fig. 1.6, however, the axes of a reference system are labeled according to some 'point of view'. According to Miller and Johnson-Laird (1976) and Levelt (1984, 1989, 1996) there are three possible 'points of view' in a reference situation. Let us consider the situation in which a ball is located with respect to a house. The ball can, for example, be *in front of the house* from the point of view of the house (which means that the ball is e.g. in the front garden). Miller and Johnson-Laird call this **intrinsic reference**. Intrinsic reference is based on an intrinsic reference frame. That is, the axes of the axial system specifying the relevant directions in this reference situation are labeled on the basis of the intrinsic properties of the Ground object (e.g. a front–back axis is available because one part of the house is recognizable as its front, and the opposite part of the house is recognizable as its back). However, the ball can also be *in front of the house* from the point of view of a speaker or a listener (which means that the ball can be at some other location than the intrinsic front of the house; e.g. at the side of the house, if the speaker or listener is facing the side of the house). In this case Miller and Johnson-Laird speak of **deictic reference**. For deictic reference the axes of the axial system used to represent the relevant directions are labeled from the viewpoint of the speaker or the listener. Finally, it is also possible to say that *the ball is to the North of the house*. In this case the system of axes that is used to locate the ball is based on relatively stable environmental features. This kind of reference is an example of **absolute reference**, and the reference frame involved is absolute in the sense that it is not dependent on some object or person present in the scene. There are thus three possible reference frames that allow us to interpret the direction in which a car drives, the location of an object, or the direction in which a weather vane is oriented. All three reference frames allow for a different way of labeling the directional axes involved (the fact that in English deictic and intrinsic reference use the same labels, i.e. top–bottom, front–back, and left–right, is not a necessary property of reference frames).

Other authors have defined different kinds of reference frames, based on slightly different properties. On the basis of an overview of the literature on reference frames, Levinson (1996) adopts a distinction between **intrinsic, relative**, and **absolute reference**. The distinction between intrinsic and relative reference is motivated—among other things—on the basis of the number of relations involved in the description. Intrinsic reference is based on the binary relation between the Figure object and the Ground object, where the Figure is located in a particular direction relative to the Ground object, based on the properties of this Ground object (e.g. *The ball is in front of me*; Levinson, 1996: 137). Relative reference is based on the ternary relationship between the viewer, the Figure object and the Ground object. Here the Figure is located in a particular direction relative to the Ground object, in terms of axes located on the Ground object, that are labeled on the basis of the properties of the viewer (e.g. *The ball is to the right of the lamp, from your point of view*; Levinson, 1996: 137). By interpreting the linguistic example *The ball is in front of me* as an example of intrinsic reference Levinson differs from Levelt (1984: 48–9), who interprets this linguistic expression as an example of deictic reference—as opposed to intrinsic reference—because the location of the ball is interpreted in relation to the deictic centre, namely the speaker. However, although the terminology applied may differ in exceptional instances, there does not appear to be substantial disagreement as to the idea that there are three different possible reference frames playing a role in linguistic directional reference.[3]

In Ch. 12 Eshuis uses the terminology relating to reference frames slightly differently, thereby refining existing distinctions. Eshuis distinguishes between classes of reference frames on the basis of (1) whether directions are extracted from the Ground object, or whether directions are projected onto the Ground object (respectively intrinsic versus relative reference) and (2) whether the source of the directional information is the ego, another object, or the environment (respectively ego-centered, object-centered, and environment-centered). By specifying different possible **origins of a reference frame** it is possible, for example, to distinguish between intrinsic object-centered, intrinsic ego-centered, and intrinsic environment-centered. These distinctions allow us to distinguish between the following linguistic examples: *The ball is in front of the house* (from the perspective of the house), *The ball is in front of me* (only my perspective is involved), and *The ball is drifting to the left* (e.g. from the perspective of the main current involved) (see Eshuis, forthcoming, for a more elaborate version of this idea).

Levelt (1996) and Levinson (1996) observe that different languages employ the three possible types of reference to varying extents. A language such as English uses all three types. Some languages, such as Tzeltal (a native meso-American language), employ two types: absolute and intrinsic reference. Other languages draw only on one type of reference frame, e.g. speakers of Guugu Yimithirr (a native Australian language) only use absolute reference. It is not possible to have a language that does not employ any kind of reference frame: In order for directional linguistic reference to take place a perceived or imagined direction must always be interpreted in relation to a set of labels that are based on a particular perspective. This crucial fact means that each of the spatial primitives assumed—vectors, axes, half-lines, and topological distinctions—must always be related to a reference frame for linguistic reference to take place. Let us consider one example here.

---

[3] In fact, Miller and Johnson-Laird (1976) and Levelt (1989, 1996) also appear to reserve the term 'intrinsic reference' for a binary relationship, and the term 'deictic reference' for a ternary relationship, thus excluding a categorization of *The ball is in front of me* as an example of deictic reference, and keeping open the possibility of categorizing it as an example of intrinsic reference. There does not seem to be a substantial difference between Miller and Johnson-Laird's (1976) and Levelt's (1989, 1996) use of the term 'deictic reference' and Levinson's (1996) use of the term 'relative reference', or in the use of the term 'intrinsic reference' by all parties. The application of the term deictic in Levelt (1984) thus seems exceptional.

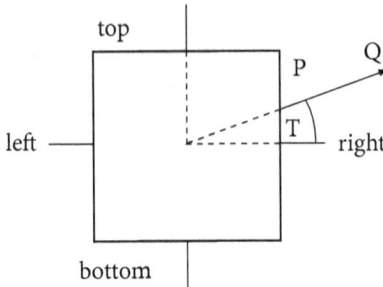

FIG. 1.7 Vector PQ is related to one axis of an intrinsic reference frame, by specifying the angle (T) the vector makes with this axis. This makes it possible to interpret the head of vector PQ as being located 'to the right' of the square.

Figure 1.7 illustrates a vector (PQ). This vector is related to at least one of two labeled axes (or a half axis), so that linguistic directional reference can take place. The fact that the axes in Fig. 1.7 happen to be labeled as a top–bottom axis and a left–right axis, and the fact that the axes happen to be located on a reference object, is not crucial. Different labels might have been used, and instead of an intrinsic reference frame some other kind of reference frame might have been used. In their chapters, Carlson, Regier, and Covey, Eshuis, and also Zwarts explicitly relate the direction vectors, used in their models to explain directional linguistic reference, to an intrinsic reference frame as shown in Fig. 1.7. In their models they explicitly relate a vector encoding direction and distance to at least one of the two intrinsic axes depicted to interpret their vectors as encoding a particular direction and distance from a Ground object.

In Ch. 11, van der Zee and Eshuis focus on the concept of an intrinsic reference frame by exploring how such frames are oriented in relation to the shape and geometry of the Ground objects on which they are centered. In a number of experiments they study the use of Dutch directional nouns and prepositions to determine the role of the geometrical features of an object in determining an object's 'front' and 'back', or the regions 'in front of' it and 'behind' it. Their experiments provide evidence for the influence of a number of spatial features, including

- the relative lengths of the intrinsic axes of the Ground object;
- the orthogonality of an axis to a curved plane of symmetry defined in the Ground object; and
- the expansion of one of the Ground object's contours along its main axis.

Van der Zee and Eshuis explain their findings in terms of a system for directional axis marking similar to that employed by Tversky (1996; Ch. 7), and Nikanne (Ch. 10). The top–bottom axis is the most directionally marked, the front–back axis is intermediately marked, and the left–right axis is least marked. Their data support this same order in the sense that, for example, main axis length 'contributes more to a directional marking of the top–down axis than to the directional marking of the front–back axis, and more to the marking of the front–back axis than to a marking of the left–right axis'. Their **Spatial Feature Categorization (SFC) model** ties all the relevant distinctions together, explaining intrinsic directional reference on the basis of the spatial features describing a Ground object. The ordering is interesting in the way it ties in with Tversky's work on verbal memory for spatial relations, and Nikanne's linguistic data on Finnish postpositions. These lines of convergent evidence for directionality marking highlight the value of a Cognitive Science approach to the area of Language and Space.

Eshuis, in Ch. 12, explicitly addresses the relation between linguistic and spatial information representation. It is argued by Hayward and Tarr (1995) and by Munnich, Landau, and Dosher (2001) that the relation between linguistic and spatial information representation can be described as a **one-to-one mapping**. These authors argue that our memory for object location is best at those locations around an object that are also occupied by the intrinsic object axes that play a role in linguistic reference (e.g. the top–bottom and left–right axis). This position is contested by Crawford, Regier, and Huttenlocher (2000), who argue that our memory for object location may be good at those axes, but is biased away from such intrinsic object axes and towards a set of intermediate axes. In line with the literature on categorization that predicts a bias away from category boundaries, and towards prototypes (Harnad, 1987), the intermediate axes are interpreted as 'prototypes' in object location memory. Since there is no direct correspondence between the prototypes in spatial memory and the prototypes in linguistic reference, it is not possible to argue for a one-to-one correspondence between linguistic and spatial information representation: 'the prototypes of linguistic categories ... are boundaries in non-linguistic spatial categorization' (Crawford, Regier, and Huttenlocher, 2000: 209).

Eshuis presents a set of experiments in which memory for object location is tested in relation to stimulus sets used by Hayward and Tarr (1995), Crawford, Regier, and Huttenlocher (2000), and van der Zee and Eshuis (Ch. 11). The linguistic categorization of the stimuli used by Eshuis is thus known. Eshuis shows, on the one hand, that the direction of the bias for object location memory can be explained by the symmetry aspects of the stimuli involved, as suggested by Tversky and Schiano (1989) and Schiano and Tversky (1992). On the other hand, he shows that the idea of interpreting linguistic prototypes as boundaries in spatial object location is only part of the story. It appears that regional errors in object location occur across such boundaries, while preserving fine-grained location accuracy. However, regional errors never 'cross over' to an opposite quadrant. It appears, therefore, that the axes that are interpreted as boundaries by Crawford, Regier, and Huttenlocher do have a role to play in non-linguistic spatial memory tasks. Given such a role, it is too simple to say that there is no one-to-one relation between spatial and linguistic information representation. The mapping may be slightly different to that expected.

In Ch. 6, Carlson, Regier, and Covey take issue with the idea of a spatial template. According to Logan and Sadler (1996) spatial templates are two- or three-dimensional fields that represent the degree to which objects appearing at each point in space are acceptable examples of such spatial relations as *above*, or *in front of* (Logan and Sadler, 1996: 496). Spatial templates are centered on a Ground object, and give different regions of acceptability for a particular spatial relation, ranging from 'good' to 'acceptable' and 'unacceptable'. Carlson, Regier, and Covey argue that the axial primitives, underlying Logan and Sadler's idea of spatial templates, are not adequate for explaining the fuzzy mappings involved in using such directional terms as *above*. Instead of distinguishing between three regions of acceptability, people seem more inclined to accept a continuum between 'good' and 'unacceptable' (see also O'Keefe, Ch. 4). In their **Attentional Vector Sum (AVS) model** Regier and Carlson (forthcoming) consider a range of data showing that the spatial relation *above* can be applied to configurations varying over a number of factors, including orientation and distance. Carson, Regier, and Covey claim that, although a reference frame is necessary to represent direction locally for intrinsic directional reference, vector representations weighing the contribution of attention and spatial factors in the reference situation play an important role in encoding the fuzzy character of the spatial relation involved.

The psychological factors that influence the fuzzy region-to-language mapping are also explored by Coventry in Ch. 13. Using spatial template theory as a starting point, Coventry shows through a range of experiments that the theory cannot account for the use of spatial prepositions without incorporating additional information relating to functional aspects of the situation being described. By considering a wider range of prepositions, including topological and directional types, he demonstrates that forms of cognitive representations underlying the fuzzy mapping are likely to involve both 'semantic' and 'pragmatic' components.

In Ch. 7, Tversky explores a different aspect of the language to space interface by looking at how people express direction and describe routes in answering a question such as *Where is X?* Assuming that people do not provide pre-stored verbal responses, they must access their knowledge of the layout of the relevant area to construct an appropriate location description. The data she examines cover both the generation and comprehension of location expressions. Her analysis of these data suggests that people will, whenever possible, make use of 'landmarks' to describe locations and layouts, and eschew an explicit specification of direction and distance. She suggests that the latter may be the case because direction and distance seem difficult to compute—when people do use direction they express it in terms of spatial relations rather than fixed bearings or angles. Tversky's findings do not seem to be consistent with a vector-based representation underlying spatial cognition and language. If the language faculty had direct access to vector representations, encoding both distance and direction, then location expressions should be replete with precise distance and direction specifications. However, this does not imply that people do not compute such variables as distance and direction. It is quite possible that some specific aspects of spatial cognition require quite precise determination of both direction and distance variables, but Tversky's evidence suggests that such representational outputs are not accessible by the language systems subserving location descriptions.

O'Keefe's work, based on animal studies, provides a wealth of evidence to support the idea that vector-based representations are employed in spatial cognition. In Ch. 4, O'Keefe extrapolates from data and computational ideas related to the spatial coding of cells in—or linked to—the rat's hippocampus to the use of spatial prepositions in human language. As a neuroscientist, O'Keefe is primarily interested in how spatial knowledge is encoded in the brain, and in particular the hippocampus. But, his work also speaks directly to the issue of how different spatial relations are encoded within a vector-based representation called a *cognitive map*, which is an absolute or allocentric spatial representation of the environment. For an understanding of spatial language, O'Keefe proposes that in humans the hippocampal representations on the left side of the brain have evolved to process linguistic and episodic rather than spatial information. This idea supports the view shared by many linguists that spatial language use, and metaphorical extensions of such language use, has a spatial representational basis. In his paper, O'Keefe introduces the **Boundary Vector Cell model**. This model assumes that place cells—representing certain locations in space—take their input from Boundary Vector Cells (BVCs). BVCs 'fire as a function of the distance of the animal in a specific direction from a large environmental feature such as the wall of the holding box'. O'Keefe identifies the representations of the place cells with the referential properties of spatial prepositions. In distinction to previous suggestions (O'Keefe, 1996), O'Keefe shows how the properties of the BVCs and place cells cooperate to form graded regions; regions that are not isotropic, but represent a continuous acceptability gradient for many different spatial prepositions.

We began this chapter by discussing the interface between linguistic and spatial information representation. Let us go back to the mappings we focused on to put the contents of this book more firmly into perspective. In the first place, the mappings depicted in Fig. 1.2 are far more complex than originally indicated. The mappings assume that only spatial features are important in determining the spatial representation of directionality. As indicated above, in relation to the work of Coventry (Ch. 13), the function or use of a particular object may also have an influence on the use of spatial language in general, and directional expressions in particular. This issue deserves separate attention, and forms the theme of the next volume in the Language and Space series.

Another aspect of the mappings in Fig. 1.2 that requires comment is that the approaches in the book cannot hope to cover the full range of possible mappings, and has necessarily to be selective. For example, the book is selective in that no chapter explicitly addresses the time factor, and that most assume a classical point-based representation of space. However, although selective, the book considers quite a wide variety of syntactic constructions and languages. The different contributions focus on the way direction is encoded in verbs, prepositions, postpositions, nouns, and adverbs, or by phrases and clauses containing these syntactic categories. In terms of languages, Nikanne shows how directionality is encoded in Finnish postpositions, while Bohnemeyer discusses directionality mainly in relation to Dutch and Yukatec. Gambarotto and Muller, Schmidtke *et al.*, and van der Zee and Eshuis and Zwarts focus—respectively—on the representation of directionality in French, German, and Dutch. Other authors, such as Carlson, Regier, and Covey, Coventry, Landau, O'Keefe, and Tversky mainly consider the encoding of directionality in English. This diversity in the coverage of $L_{directional}$ also contributes to the inventory of language-specific constraints on mapping directional spatial representations to $L_{directional}$, and vice versa.

Finally, the authors in this book focus on the mappings depicted in Fig. 1.2 in different ways. This follows from the fact that the authors belong to different disciplines in the Cognitive Science society: Cognitive Anthropology, Cognitive Neuroscience, Cognitive Psychology, Computer Science and Linguistics. For example, linguists, and in particular logical semanticists, try to identify a formal semantic model for the set of expressions that comprise $L_{directional}$. Such a model contains all the elements assumed to specify the meaning of $L_{directional}$ by deriving a formal correspondence between the elements of the model and the spatial primitives assumed to encode directionality. In contrast, psychologists tend to focus on the use of $L_{directional}$. Their goal is to investigate the cognitive structures and processes that need to be imputed to account for the behavioral data derived from their experiments. They are interested not just in the way $L_{directional}$ corresponds to cognitively specified directional primitives, but also in how the cognitive systems underlying these domains relate to other cognitive systems, such as our more general capacity for thinking. Computer scientists, on the other hand, tend to put the emphasis on the formal representation of space and direction, making specific assumptions about the nature of the spatial primitives involved, and how these primitives relate to $L_{directional}$, or other direction encoding. The Cognitive Neuropsychological work of O'Keefe, and the Cognitive Anthropological work of Bohnemeyer draw on many of the distinctions discussed.

With the exception of the first two, the order of the chapters is loosely structured around the richness of the spatial representations the different authors employ to encode direction. Chapter 2, by Landau, advocates an axis-based approach, and provides a natural contrast with the vector approach advocated by Zwarts in Ch. 3. Following Zwarts's chapter, Chs. 4, 5, and 6 all assume vector representations for encoding direction. Ch. 7 presents findings that do not seem consistent with a vector-based approach. Chs. 8 and 9 demonstrate that more minimal representational machinery is sufficient for representing direction. Chs. 10 and 11 adopt the slightly richer axis approach to represent direction. Ch. 12 combines an axis and a vector approach, whereas Ch. 13 comments upon the use of geometry-based approaches in general, and shows that direction representation also requires looking at the function of the objects involved. Taken together, the chapters provide an overview of current ideas and positions in relation to the formal or cognitive encoding of direction for language.

2

Axes and Direction in Spatial Language and Spatial Cognition

BARBARA LANDAU

Abstract

This chapter argues that axial representations are engaged in both linguistic and non-linguistic tasks. Axial structure is required to account for performance in object location memory tasks, matching tasks, and spatial language tasks. Axial structure representations are independent of direction representation, as revealed by studies of normal adults and children, and spatially impaired adults and children. More specifically, evidence from spatial impairment suggests that direction may be a more fragile component of spatial representation than axial structure. The chapter concludes by arguing that in relation to spatial language axial representations are more suitable for representing direction than vector-based representations.

Introduction

Spatial cognition—knowledge of objects, motions, and spatial relationships—is one of the most fundamental functions of the human mind and brain, emerging early in development with no formal tutoring, and supporting a wide range of activities from navigation to map use to problem-solving. The crowning touch of human spatial cognition is our capacity to talk about space. All human languages have rich resources for encoding spatial entities, including objects, events, and spatial relationships. This capacity is widely assumed to build upon non-linguistic representations of space. For example, H. Clark (1973) proposed

---

The work reported herein was supported in part by grant #12FY98-194 and 12FY99-670 from the March of Dimes, grant 1 R55 NS37923 and B CS 0117744 from the National Institutes of Health, and grant SBR 9808585 from the National Science Foundation.

that spatial terms such as *above, below, left,* and *right* are supported by our non-linguistic representations of these relationships, with priority given to the terms that refer to vertical (gravitational) axes, relative to those referring to the horizontal, or non-gravitational axes. E. Clark (1973; 1980) argued that the early acquisition of spatial terms can be predicted from children's non-linguistic biases: terms whose meaning meshes with non-linguistic biases should be easiest to acquire. Other theorists have similarly argued that non-linguistic spatial representations should play an important causal role in shaping spatial language. For example, O'Keefe and Nadel (1978) first proposed that language might have evolved as an encoding of non-linguistic representations of space, specifically that the left hippocampus in humans might preferentially encode those spatial semantic primitives that are derived from non-linguistic primitives encoded in the right hippocampus for other spatial purposes. More recently, Landau and Jackendoff (1993) proposed that the geometric properties captured by the closed class set of spatial terms might correspond to the 'where' system of the brain, as described by neuroscientists such as Ungerleider and Mishkin (1982). The *basic* spatial terms—those that are monomorphemic—are few in number, and they appear to capture a distinctive set of spatial relationships. In Landau and Jackendoff's scheme, these spatial 'meanings' might have originated in the way that the brain encodes location non-linguistically.

Each of these proposals is based on the assumption that spatial language has its roots in non-linguistic spatial representations. If this is true, then an analysis of spatial language can offer us considerable insight into the nature of non-linguistic spatial representations (O'Keefe and Nadel, 1978; Landau and Jackendoff, 1993).

The purpose of this chapter is to use one small group of the closed class spatial terms to explore the primitives that might be shared by both spatial language and non-linguistic spatial cognition. In accordance with the theme of the volume, I will concentrate on terms that appear to engage representations of **axial systems**. Such representations can explain systematic effects showing that when we represent an object's location, we do so in terms of a **set of orthogonal axes** with its **origin** centered on the **Reference object**. However, I will also argue that the axial system must be enriched by the representation of **direction** *within* each axis. Language provides the principal evidence for this component, but non-linguistic evidence points to the same conclusion. This axial and directional representation is somewhat different from the **vector representation** proposed by O'Keefe (1996; Ch. 4), even though both represent direction. I will discuss several differences between the proposals at the end of the chapter.

An axial system that includes specification of direction can adequately capture how children and adults produce and understand a small but key set of basic spatial terms. Such a representation also seems to capture certain facts about non-linguistic spatial representations. Furthermore, empirical evidence suggests that in both systems, the axial system may be represented in the brain *separately* from the system that specifies direction within axes. To illustrate this, I will summarize evidence from individuals with spatial impairment, as well as evidence from normal children and adults.

In the literature on spatial cognition and spatial language, there is considerable confusion, unclarity, and variability in terminology regarding axial systems. In this chapter, I use **axial system** to refer to a set of orthogonal axes, which will typically have its origin centered on some Reference object. Reference objects can be found in aspects of the environment, objects in the environment, the body itself, or other locations that are the focus of attention (McCloskey and Rapp, 2000). In the literature, axial systems are often referred to as **reference systems**, although actually, reference systems need not be axial in nature. Since 'reference system' usually means 'axial system' in the literature, I will use these terms interchangeably. Axial systems can be two-dimensional (two axes) or three-dimensional (three axes). If we think of them like **coordinate systems**, they will include metric values and direction. For the purposes of the *basic* spatial terms I will consider, metric values do not seem to be necessary, but direction definitely is necessary. As O'Keefe (1996; Ch. 4) points out, language certainly can encode metric values of distance (e.g. *John walked straight ahead for three miles*); but this aspect of the spatial representation is encoded by modifying phrases, and hence may not constitute the most basic aspects of space that are encoded by language. For the purposes of some non-linguistic tasks, *both* metric values and direction seem to be necessary. For example, when we navigate through space, we represent the locations of places relative to each other using both direction and distance (Gallistel, 1990; Philbeck, Loomis, and Beall 1997).

In the sections that follow, I will first review evidence that language engages representations of multiple axial systems as well as direction. Then I will review evidence that axes and direction are represented similarly—though not identically—in linguistic and non-linguistic tasks tapping spatial representation. Following this, I will review evidence showing that axes and direction may be represented *separately* in both language and non-linguistic spatial systems. Finally, I will present some thoughts on how a system of axes plus direction (which captures a variety of data quite well) might differ from a vector system of representation such as the one offered by O'Keefe.

1. Multiple Axial Systems in Spatial Cognition and Spatial Language

There is abundant evidence that the brain and mind represent space in terms of multiple reference systems. As I have already indicated, locations are usually represented in terms of a system of axes whose origin is centered on some Reference object. Hence if the Reference object is the body, then the origin would be centered on the body; if the Reference object is the head, the origin is centered on the head, etc.

Studies show that both humans and other animals represent location in a variety of different reference systems, that is, locations can be specified relative to a variety of different Reference objects, or origins situated on those objects. For example, studies of alert monkeys performing spatial tasks show that certain neurons in one area of the parietal cortex respond selectively to stimuli in receptive fields defined by **head-centered axes**, whereas neurons in an adjacent area of the brain respond selectively to stimuli in receptive fields defined by **eye-centered axes** (Colby and Goldberg, 1999). Studies of animals moving through space to reach targets indicate that they represent location in terms of distance and direction from key aspects of the environment (O'Keefe and Nadel, 1978; Gallistel, 1990), indicating that they use reference systems centered on aspects of the environment. If deprived of rich information about the environmental layout, they can also use body-centered reference systems, as shown by the large body of research carried out during the early and mid-twentieth century (see O'Keefe and Nadel, 1978; Gallistel, 1990 for reviews).

Studies in human neuropsychology also reveal that the brain represents location in terms of multiple reference frames (see e.g. Carlson-Radvansky and Jiang, 1998; Logan, 1995). Much of the pertinent evidence comes from the study of neglect, a syndrome in which the patient ignores, or fails to acknowledge the presence of stimuli in the region of space contralateral to the lesion. **Neglect patients** often show lesions in the parietal lobe, especially in the right hemisphere (see Behrmann, 1999, for review). Thus the patients often neglect the 'left' region of space. In one classic task used to diagnose neglect, patients are shown a sheet of paper with a large set of randomly oriented lines. When they are asked to cross out all the lines, patients often cross out only the lines on one side of the paper, e.g. the right side, leaving the lines on the left side intact. Thus, the patient is said to neglect the left side. But of course, the notion LEFT must be defined relative to a reference frame in order to be meaningful: is the region LEFT relative to a retina-centered reference system, a head-centered system, a body-centered system, an object-centered system, an environment-centered system, and so forth?

Evidence shows that neglect can occur for a variety of reference systems. For example, Bisiach and colleagues (Bisiach, 1996) described a neglect patient who was asked to imagine himself in a salient location in his home town (e.g. on the steps of a church in the center of a piazza), and then to describe the rest of the piazza. He described only the buildings and landmarks that were on one side of the piazza, the side contralateral to the lesion. Then he was asked to imagine himself at a location at the other end of the piazza, facing the opposite direction, towards the original location and he was again asked to describe the piazza. From this location, all buildings originally to his right would now be to his left in the layout, and vice versa. The patient now tended to describe the buildings and landmarks that he had neglected in the first condition; that is, he again described the layout as imagined from the 'left' side, defined in an egocentric reference system. These striking observations show, first, that neglect affects not only the current perception of space, but remembered space, and second, that it can affect locations defined in a **body-centered reference system**.

Other studies have shown that neglect can also affect **object-centered representations**. For example, Tipper and Behrmann (1996) asked neglect patients to detect a flashing light presented at one of two ends of a single object (a dumb-bell) that was presented in the center of their visual field. Patients showed the typical neglect for the side contralateral to their lesion. Tipper and Behrmann then rotated the dumb-bell around the center point, with the result that the object's left side moved to the right visual field, and vice versa. Patients continued to neglect the side of the object that they had originally neglected, even though it was not in the visual field contralateral to the lesion. This result suggests that the patients were neglecting one side of the object—say, the 'left' side—as defined in an object-centered reference system.

Thus, the literature on systems other than language indicate that reference frames—sets of orthogonal axes with an origin—serve as organizers in many domains. These include the retina, eye, head, body, object, and environment.

What about language? Languages have the resources to encode location relative to any of the reference systems described. For example, one can describe a location as *to the left*, in a **retinocentric reference system**. However, words such as 'retinocentric' are technical, specialized terms used in restrictive settings. The basic terms of a language—those that are morphologically simple, widely used, and learned early in life—do *not* include *separate* terms that specially mark each reference system. Thus, for example, we can talk about locations in retina-centered, head-centered, or body-centered reference systems, but in each case, we refer to locations within these reference systems using the same basic set of terms, e.g. *above, below, right, left*. That is, the same basic set of terms can be used to describe location over a number of reference systems, without further special marking.

This might suggest that languages do not mark reference systems at all. However, this is false. There are two clear exceptions in English and other languages. First, locations defined in an **object-centered system** draw on a special set of terms, the *top, bottom, front, back,* and *side* of objects. These terms are used to describe locations within objects, not between objects. For example, if we wish to refer to the *top* of a sugar bowl, this region is the same regardless of the bowl's location or orientation in space. Similarly, the *bottom* is usually at the opposite end of the main axis of the object from the *top*. Terms marking these object-centered regions are seen across a variety of languages, and are sometimes derived from the names of body parts (e.g. in Tzeltal and related languages, see Levinson, 1992*b*). To do so, as with English terms *top, bottom,* etc., the application of the terms appears to be guided by perceptual representations of the object's shape, in particular its main and subsidiary axes. In this case, the spatial language system must interact with another spatial system, in this case the object perception system.

A different set of terms specially marks location relative to **the earth's coordinate system**. In English, these terms are *north, south east, west*; many other languages specially mark these locations as well. Some languages also have terms describing location relative to local geographic features, e.g. *uphill, downhill,* or *seaward* (though the latter tend to be morphologically complex, i.e. compounds).

Thus, languages appear to engage representations of a variety of reference systems, as do other spatial cognitive functions. However, locational terms in natural languages do not specially mark most reference systems, even though these must be differentiated for the purposes of action (e.g. a location defined as ABOVE in retinocentric terms may be quite different from ABOVE in an object-centered system).

2. Empirical Evidence for Axial Representations in Spatial Language and Spatial Cognition

We will now consider a small set of basic terms that describe the location of objects relative to a variety of reference systems. This set includes terms *above, below, left,* and *right*. These terms encode the location of a select set of regions outside an object (i.e. they are not used to refer to spatial regions *within* an object). The regions appear to be defined with respect to the axes of the Reference object (suitably modulated by attention, see Carlson, Regier, and Covey,

Ch. 6). A typical complete expression of location in English will include an NP expressing the object to be located (the **Figure object**), an NP expressing the object in terms of which the Figure is located (the **Reference object**), and a spatial preposition, which expresses a function on the Reference object. Thus, in the expression *the X is above the Y*, the X is the Figure, the Y is the Reference object, and *above* specifies a region relative to Y, within which X is located.

A number of researchers have asked whether the representations underlying these spatial terms are specific to language or are shared across linguistic and non-linguistic spatial representations. In order to find the answer, one needs tasks that can be carried out in both a linguistic and a non-linguistic mode, using identical materials and procedures that are as similar as possible. These criteria were first met by Hayward and Tarr (1995), who carried out two tasks with native adult speakers of English. In the language task, Hayward and Tarr showed people displays of a Figure object in various locations around a square Reference object, and asked them to fill in the blank in the sentence, *The X is ____ the Y* (where X was the name of the Figure object and Y a square Reference object). Looking only at uses of basic (monomorphemic) spatial terms, Hayward and Tarr found that people used these terms most densely along the extended axes of the Reference object. For example, the term *above* was used most densely along the vertical 'positive' axis (i.e. the one extending upwards from the Reference object); *below* or *under* was used most densely along the vertical 'negative' axis (i.e. the one extending downwards from the Reference object); and *left* and *right* were used most densely along the horizontal axis extending in the appropriate direction from the Reference object. In a memory task, people were shown briefly presented displays showing Figure objects located either on the extension of the Reference object's axes, or off that axis. After each display disappeared, a visual mask was presented briefly, followed by a second display that showed the Figure object either in the same location as the original, or in a different location, displaced by a small amount from the original. Hayward and Tarr found that people's memory for location of the Figure object was more accurate for targets that fell along the Reference object's axes than those off-axes, again indicating that axial structure played an important role in people's representations of location (but see Crawford, Regier, and Huttenlocher, 2000, for a different interpretation).

These results suggest that the structures underlying spatial language and spatial cognition (here, memory) both engage axial representations, in which people mentally impose an axial structure on the Reference object, and represent the Figure object's location in terms of those axes. Recently, Munnich, Dosher, and Landau (2001) carried out a modified replication of Hayward and Tarr's study, comparing adult native speakers of English and adult native speakers of Japanese (in which the meanings of basic spatial terms differ somewhat from English). They found the same pattern of dense application of basic spatial terms along axes across the two languages. Moreover, they found the same pattern of dense accuracy for locations that fell on the Reference object's axes, relative to those that did not. These findings suggest that the axial representations uncovered by Hayward and Tarr may be universal, and in any event are not a necessary consequence of having learned English. A schematic illustration of Munnich et al.'s findings are shown in Fig. 2.1.

These results suggest that a partial homomorphism exists between spatial language and spatial cognition. I say 'partial' because the findings also indicate that the structures are somewhat different in the two systems. For example, the results of Munnich et al.'s memory task showed *graded* effects of distance,

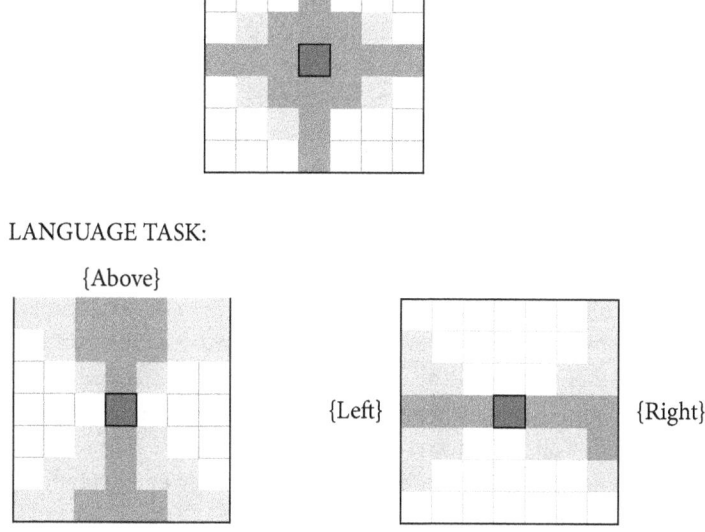

FIG. 2.1 Schema of results for two spatial tasks (after Hayward and Tarr, 1995; Munnich, Landau, and Dosher 2001). In the Memory task, people remembered best the locations along the extension of the Reference object's axis (better performance shown by darker image). In the Language task, people used basic spatial terms most densely along the same axes (darker image). In addition, the Language task showed effects of direction, with people distinguishing between 'positive' and 'negative' ends of the vertical and horizontal axes (above–below and left–right).

with accuracy declining smoothly as the location moves further away from the Reference object's axis. However, the language task showed categorical effects of distance, with use of terms falling off very sharply just outside of the axis itself. Similarly, there were graded effects of distance *within* an axis in the memory task, with accuracy declining with locations farther from the Reference object, but no such effects in the language task. These findings show that language uses categorical distinctions, which are blind to detailed effects of distance. Memory appears to be more sensitive to distance. Some investigators have suggested that there is even less to the idea of a homomorphism, since some memory tasks show categorical effects that are not seen at all in language tasks (Crawford, Regier, and Huttenlocher, 2000). It certainly seems possible that, depending on the nature of the memory task, one might find very different kinds of structures from those normally reflected by language. However, it is also the case that Hayward and Tarr, Munnich et al., and even Crawford et al. have found substantial similarity—not identity—across tasks tapping linguistic and non-linguistic spatial representations.

In sum, these findings clearly indicate that spatial language engages an axial system that is in some ways quite similar to that tapped by non-linguistic spatial memory. In addition, the results of the *language* tasks show that, within the axial system, *direction* is represented. This is required in order to explain people's application of terms *above* vs. *below*, and *left* vs. *right*. Note that directional effects were not uncovered in any of the memory tasks I have discussed, but this does not mean that they do not exist. The experiments were not designed to test the representation of direction, which would have required including distractors that were identical in location in every way *except for* direction. Such a test might include, for example, a target item on-axis to the right of the Reference object, and test items that were either the same or different in direction only—that is, on-axis at the same distance from the original target, but now to the left. There is some pertinent evidence on this distinction between axes and direction within axes. This evidence suggests that axes and direction are represented separately in both spatial language and spatial cognition. We turn now to this evidence.

3. Evidence for the Separate Representation of Axes and Direction in Spatial Language and Spatial Cognition

Language production data among adults already show that a small set of spatial terms maps onto axial representations together with direction. The evidence described so far does not indicate whether both axes and direction are encoded in non-linguistic spatial tasks. Moreover, the fact that language encodes both axes and direction raises the question of whether these are bound together and indissociable in language. If they are dissociable in language, one might ask whether they are also dissociable in non-linguistic tasks.

There are several lines of evidence suggesting that direction may be represented *separately* from axes in both spatial language and spatial cognition. In general, this evidence shows that one can eliminate or impair directional coding while not eliminating or impairing axial representation.

One source of evidence comes from the representation of spatial language and spatial cognition in children who have **Williams syndrome (WS)**, a rare genetic defect characterized by a micro-deletion on chromosome 7. The deleted genes include ones for elastin, LIMK1, and a number of others (Morris *et al.*, 1994). The link between the particular genes deleted and the WS phenotypic profile is not well understood, but some have speculated that LIMK1 is specially linked to part of the unique and characteristic cognitive profile found in people with WS.

The phenotypic profile of individuals with WS includes distinctive facial characteristics, and various medical problems often linked to the missing gene for elastin (which has ubiquitous effects in the body). Of particular interest to cognitive scientists, individuals with WS have a distinctive *cognitive* profile, including profoundly impaired spatial cognition together with relatively spared language (Bellugi *et al.*, 1988, 1992). Their spatial impairment has been most frequently noted in the context of visual-spatial construction tasks, which are often used as part of general intelligence test batteries. In these tasks, people are asked to copy a line drawing, or to replicate a pattern that is composed of individual blocks whose spatial arrangement results in a design. Both of these tasks are typically administered with the model in view, so any deficits cannot be ascribed solely to poor memory. Individuals with WS are profoundly deficient in these tasks, typically performing in the bottom percentiles for their age. For example, adolescents with WS typically perform at levels of 4–5-year-old children (Bellugi *et al.*, 1988; Mervis *et al.*, 1999; Hoffman, Landau, and Pagani, 2002). An example of a copy made by a WS adolescent and a normally developing 5-year-old child matched for mental age is shown in Fig. 2.2. Clearly, the overall spatial configuration is not preserved by the WS individual, whereas it is preserved by a normally developing child who is much younger. Although the nature of this spatial deficit is not yet understood, there is reason to believe that some fundamental properties of spatial representation are impaired.

Unlike the profile of spatial impairment, WS individuals have remarkably rich linguistic capacities. Their language is fluent, and generally speaking, well-formed, both syntactically and semantically. Their vocabulary tends to be quite

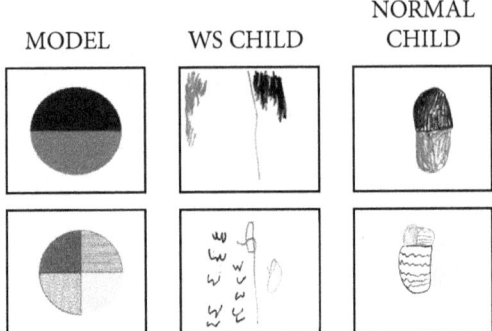

FIG. 2.2 Samples of drawings by a child with Williams syndrome (WS) and a normally developing child matched for mental age. Children copied the Model shown on the left while it was in full view. The child with WS shows the spatial disorganization typical of children with this syndrome.

good, with scores at least as high or higher than would be expected by their overall mental age (Mervis *et al.*, 1999). Scientists have been struck by the fluency of language among WS individuals, and have noted that it is especially rich in intonation and overall 'melody' (Bellugi *et al.*, 1988, 1992).

The stark contrast between language and spatial cognition in WS affords a unique opportunity to examine the link between the two systems. Simply put, if WS children have severely impaired spatial cognition, how can they learn *spatial language*, using nouns, verbs, and prepositions to refer to objects, motions, and spatial relationships? In our laboratory we have been considering just this question, by examining the nature of spatial breakdown in WS people the nature of their spatial language, and the possible links between the two systems.

In one set of experiments we used the methods developed by Hayward and Tarr (1995) and by Munnich *et al.* (2001) to examine spatial representation in a language task, and a corresponding non-linguistic task (Landau, 1999, 2003; Zukowski, Schwartz, and Landau, 1999). Children with WS between the ages of 8 and 13 were tested in a modified version of Munnich *et al.*'s methods. Because people with WS are moderately retarded, we also tested a group of normally developing children who were individually matched to the WS children on mental age, using a standardized intelligence test. As in Munnich *et al.*'s study, children were tested in linguistic and non-linguistic tasks that tapped their capacity to talk about location and to represent location non-linguistically.

In the non-linguistic task, children viewed a display showing a dot (the Figure) located somewhere around a square Reference object (as in Munnich *et al.*). Their attention was drawn to the location of the Figure relative to the Reference object, without using specific locational terms. Below this display were two test displays. One was the same as the target, and the other showed the Figure object displaced by 0.25 inch. Children were asked to point to the test display where the dot was *in just the same place to the square* as in the original display. The space around the Reference object was sampled using 36 target locations, including 12 in which the dot was in contact with the square, 8 in which it fell along the extension of the square's main axis (on-axis) but not in contact, and 16 in which it fell neither along the square's axis extension nor in contact with the square (off-axis, no contact). Our question was whether WS children would show any of the structure that normally emerges among adults in the memory task, specifically, whether axial structure would be preserved.

All children (WS and normal controls) performed better for targets that were *on-axis* than those that were *off-axis*. Thus both groups showed **representation of axial structure**. At the same time, there were some differences between the groups. First, children with WS showed less of an axial advantage (i.e. less advantage on the axis), as the target was located at greater distances from the Reference object. Second, although both groups of children matched targets better when they were on the vertical rather than horizontal axis of the Reference object, children with WS showed disproportionate fragility on the horizontal axis. These two outcomes are shown schematically in Fig. 2.3. Thus, even

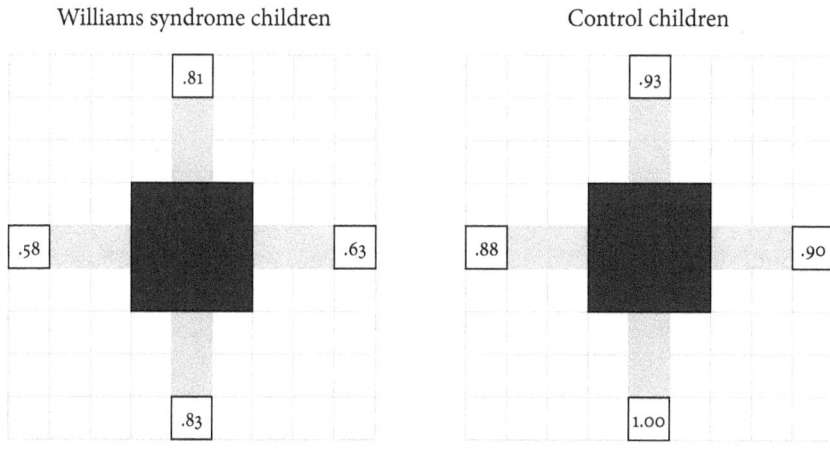

FIG. 2.3 Mean percent correct for locations that are distant from the Reference object (black square), along vertical and horizontal axes. Overall children with WS and normal controls show the same qualitative pattern of advantage on axis, with strong performance on-axis (compared to off-axis, not shown here) and advantage for the vertical axis relative to the horizontal. However, the children with WS show disproportionately worse performance along the horizontal axis.

though children with WS are severely impaired in non-linguistic spatial cognitive tasks, they do show *qualitatively* normal axial structure in this simple locational matching task. Their weakness appears to be due to a more fragile axial system, in which the axes break down over distance from the Reference object, and show particular weakness along the horizontal.

The same children were tested on two language tasks, and these again revealed considerable axial structure. In the production task, the children were shown the same target locations as in the non-linguistic task, one at a time, and asked *The dot is where to the square?* In their productions, WS children and normal children freely produced basic axial terms such as *above, below, next to, beside*, and they distributed these terms most densely along the correct axis. Figure 2.4 shows the distribution for the vertical positive terms that were produced (e.g. *above, on top of, over,* etc.). Note that the terms were produced

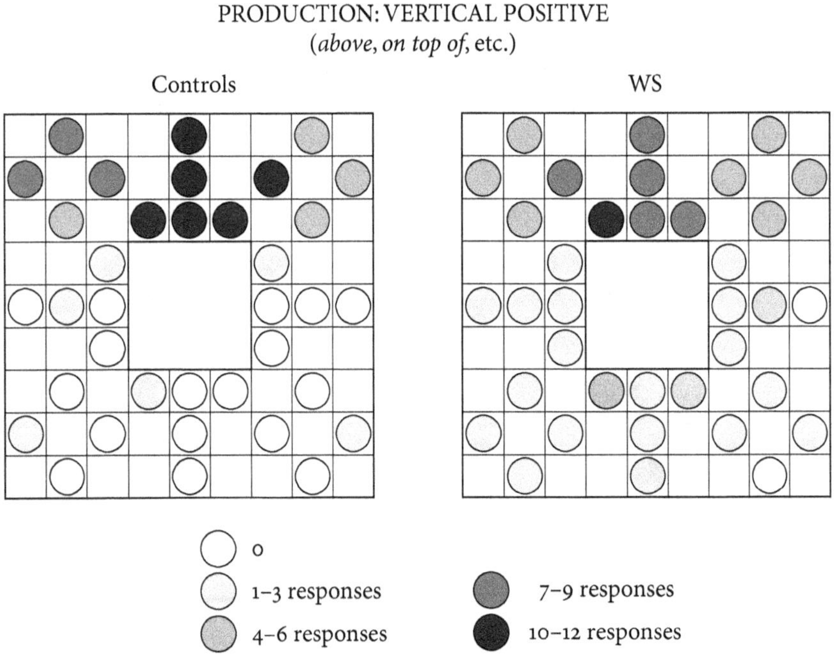

PRODUCTION: VERTICAL POSITIVE
(*above, on top of,* etc.)

Controls   WS

○ 0
◯ 1–3 responses   ● 7–9 responses
◉ 4–6 responses   ⬤ 10–12 responses

FIG. 2.4 Locations for which children produced vertical positive terms (e.g. *above, over, on, off,* etc.). The data represent production from 12 children with WS and 12 normally developing children matched for mental age. Each child named each location (circle) once; the density of color in each circle represents the number of children producing vertical positive terms for that location.

largely for locations that were, in fact, in the vertical positive region of the Reference object. Production of the vertical negative terms (not shown in the figure) was distributed in the complementary region, appropriately for these terms. And production of the horizontal terms (mostly *beside* and *next to*) was distributed most densely along the extension of the horizontal axis. Thus both vertical and horizontal axes were respected, with terms such as *above/below* being produced along the vertical axis, but not the horizontal axis; and terms such as *beside* produced along the horizontal, not the vertical axis.

Despite this normal aspect of axial representation, the WS children's productions showed some differences from the distribution of normal children's productions. In particular, the organization of WS children's productions was less tightly distributed along the axes than that of normal children. Moreover, WS children used more global terms than normal children (e.g. many used 'near') and they produced *errors of direction*, in which, e.g. *above* was used for a location below the square, or vice versa. Figure 2.4 shows that WS children, but not normal children, sometimes produced a vertical positive term for locations that were clearly *below* the square. In recent work we have observed this pattern even among adults with Williams syndrome, indicating that, although the pattern is sporadic in any single sample, it may persist over development. It is important to note that we observed this pattern only once among our normally developing children, and this was in the youngest child sampled (a 3-year-old). Complementary evidence on production of horizontal terms indicates that directional errors along this axis also persist among WS adults, with these individuals correctly producing terms *right* and *left* for locations along the horizontal axis, but incorrectly assigning these terms to the correct direction.

Children were also given a comprehension task, in order to provide converging evidence, and to test terms that were not typically produced, such as *left* and *right*. In this task, children were given a blank piece of paper showing the Reference object and were asked to *place a dot _____ the square*, using each of 14 vertical axial terms (7 'positive', e.g. *above, on top of*, etc., 7 'negative', e.g. *underneath, under*, etc.) and 8 horizontal axial terms (*right, left, next to, to the right of, to the left of, beside*, etc.). Each term was tested individually on a separate page in order to avoid spatial contrast effects. The distributions of the WS and normal children's responses preserved axial structure quite well, with dots for vertical terms (e.g. *above/on top*) along the square's vertical axis extension (as in Fig. 2.5) and dots for horizontal terms (*next to/right/left*) along its horizontal axis. Again, however, the WS children's distribution was more fragile than the controls'; and they often got the *axis correct, but direction wrong*. For example, as can be seen in Fig. 2.5, WS children, but not controls, sometimes placed a dot for the term *above* the square below it. Only one control child made such vertical axis errors—again, the youngest child we tested. Moreover, directional errors for right/left were persistent even among adults with Williams syndrome, although they tend to disappear in normal children by around age 8.

Thus the evidence on axial representation from the language tasks was consistent with the evidence from the non-linguistic task: the children with WS, like normal children, represented location in terms of axial structure organized around the Reference object. Just as with the evidence from the non-linguistic task, the language data show that the axial system is represented by WS children in much the same way as for normally developing children. However, in both the language and non-linguistic tasks, the WS children's representations of axes appears to be less tightly organized than that of normal children. The evidence from both language tasks also showed that, in some cases, children produced or comprehended spatial terms using the appropriate axis, but

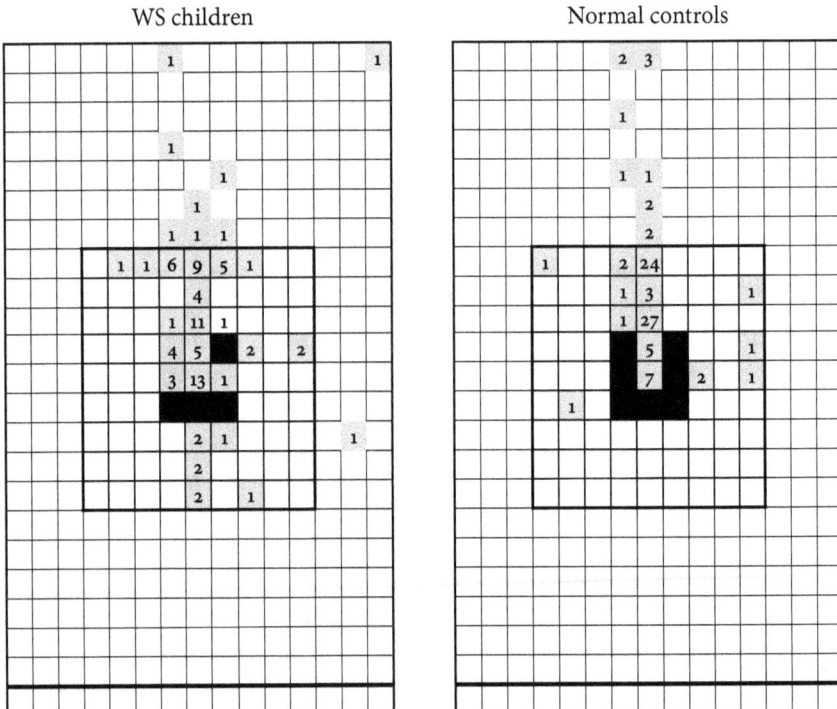

FIG. 2.5 Children's dot placement locations when asked to put a dot *above/over/on top of* the black square. The data represents comprehension by 12 children with Williams syndrome and 12 normally developing children matched for mental age.

erring on direction within the axis. Although the figures show only data for vertical terms, the evidence for horizontal terms is consistent with this picture. In particular, the children showed axial structure for horizontal terms in both the production and comprehension tasks. The comprehension task, which explicitly tested terms that encode axis and direction, such as right and left, showed that both WS children and normally developing children made errors that preserved the horizontal axis but did not preserve direction. For example, children in both groups sometimes placed a mark to the right of the Reference object when queried on the term *left*, and to the left of the Reference object when queried on the term *right*. However, even when they erred on direction, they were correct in placing marks along the extension of the horizontal axis, indicating that they had already organized the terms *left* and *right* as horizontal terms.

In sum, the data from the non-linguistic and linguistic tasks showed the existence of considerable axial structure, even among children who are severely spatially impaired. The data from the language tasks show additionally that, even when axial structure is preserved, the representation of direction may be impaired. The fact that the normal control children showed the same pattern of preserved axial structure but faulty representation of direction within axis (in the case of RIGHT and LEFT) indicates that the pattern shown by WS children may reflect delayed development, not qualitatively different kinds of representation.

The separate representation of axis and direction was discovered through tests of spatial language. Is the separation of axis and direction also a characteristic of non-linguistic spatial cognition? Several lines of evidence suggest that it is.

First, in a follow-up experiment to our non-linguistic matching task, we asked whether children and adults would show any evidence of such separate representations. We used the matching task again, but changed the distractors. This time, the two test arrays included one array that was identical to the target array, and one that differed in one of two ways. Half the different arrays showed the Figure object in a location that preserved the axis but changed direction, and the remaining half showed it in a location that changed both axis and direction. Thus, for example, if the target array showed a Figure object located along the vertical positive extension of the square's axis (i.e. *above* the square), then the 'different' array showed the Figure either (1) *below* the square, at an equal distance as the original target, or (2) to the *left* or *right* of the square, at an equal distance as the original target. In the first case, the Figure object lies along the same axis, but in a different direction; in the second case, it lies along a different axis entirely. We tested WS children, normally developing children, and normal adults in a No-Delay condition, in which both target display and test displays were viewed simultaneously, and a Delay condition, in which the target display was viewed for several seconds, then disappeared, and the test displays appeared.

All groups were at ceiling in the No-Delay condition. This is not surprising, since the 'different' displays showed the Figure object in locations that were grossly different from the target. In the Delay condition, however, overall performance dropped slightly for adults, moderately for normal children, and massively—closely to chance levels—for WS children. The patterns of reaction time among adults, and the patterns of correct performance among children all showed the same effect: distractors that lay on the same axis as the target (but differed in direction only) elicited worse performance than distractors that lay on a different axis. This pattern showed up as lower percents of correct matching among children and longer reaction times among the adults. In both cases, the pattern indicates that the axis is represented separately from direction, and that coding direction is more difficult than coding axis alone.

This finding may seem counterintuitive. But it is consistent with the data from the language tasks. It is also consistent with several findings from other, quite different literatures.[1] Logan (1995) tested adults' ability to verify whether a target matched a basic spatial term that preceded it. People were shown the term *above*, *below*, *left*, or *right*, and then shown a display in which a target object (a dot) was in the specified location or in a different location. All target dots were located along the extension of the main axes of the Reference object, either above, below, to the left, or to the right of it. The three distractor-dots were located along the same axis as the target, but in the opposite direction (e.g. *below* the Reference object for the target location *above*) and at each end of the orthogonal axis (e.g. *right* and *left* of the Reference object for the target *above*). In half the trials, the same-axis distractor was the same color as the target, and in the other half this distractor was a different color. Logan found that subjects were faster to verify relations *above* and *below*, than *left* or *right*, consistent with other literature (e.g. H. Clark, 1973, Franklin and Tversky, 1990) and with our own findings of higher accuracy along the vertical axis than the horizontal axis. In addition, however, Logan found that subjects were faster to verify the target relationship when the same-axis distractor was the same color as the target. Logan interpreted this finding to mean that subjects could separately set the two key parameters of location—axis and direction. In the case of same-color distractors located on the same axis as the target, people

---

[1] I thank Laura Carlson for suggestions regarding the separate representation of axis and direction in spatial attention tasks.

could verify the relationship by verifying the axis alone, without direction. This would be made more difficult in the case of different-color distractors on the same axis as the target, since both axis and direction would have to be checked in order to verify location.

A similar phenomenon was reported by Carlson-Radvansky and Jiang (1998), who examined whether reference systems could be negatively primed, resulting in slower reaction times when they were used. People were asked to verify spatial relationships between two objects, after seeing a spatial term (e.g. *above*) that specified the relationship. In prime displays, people saw, for example, X above Y in an environment-centered frame of reference (which would, by hypothesis, inhibit other frames of reference, e.g. an object-centered one). When people then verified the relationship on the next trial, inhibition due to priming would result in slower responses to relationships whose verification required a frame other than the one activated during the prime trial. Carlson-Radvansky and Jiang did observe such inhibition, and found that it operated along an entire axis of the reference frame. That is, there was inhibition whether the new relationship was *above* or *below* using a different frame of reference. This again indicates that the axis itself—regardless of direction within the axis—can be activated and used during cognitive processing.

Finally, studies of a spatially impaired adult have shown that some spatial breakdown can be characterized as impairment to the directional component of location, with preservation of the axial component. McCloskey *et al.* (1995) reported a case study of a college student, A.H., who has a developmental deficit in representing the location of objects. Although she generally functioned quite well, as evidenced by her successful completion of college at a highly competitive university, A.H. showed gross deficits in spatial representation. For example, when asked to copy complex figures, she often copied components from the right side of the target Figure onto the left side of her copy, and vice versa. When asked to move a computer mouse to indicate the location of an X presented on the screen, she made numerous errors, and all errors were confusions between left and right *or* between vertical positive (up) and vertical negative (down). When asked to close her eyes, then open them and immediately reach for an object, she reached to the left for targets on the right, and to the right for targets on her left. She accurately reached all targets in the center. In follow-up experiments, McCloskey and Rapp (2000) showed that A.H.'s localization errors were due to impairment in the **directional component** of a coordinate system whose origin was determined by her focus of attention. Specifically, as she focused attention on different locations, the origin of the coordinate system moved; from this origin, the axis and distance along the axis were preserved, but direction was not preserved.

Thus a variety of evidence shows that axial representations are preserved in non-linguistic tasks as well as in tasks engaging spatial language. They further show that direction is preserved in normal representation, but that location can also be represented in terms of axes without direction. The fragility of direction under spatial impairment, as well as its optionality in normal spatial representation show that it is represented separately from the axes themselves.

4. Conclusions: Axial Representations, Direction, and Vector Representations

The results I have reviewed show that axial representations are engaged in both linguistic and non-linguistic tasks. Axial structure is required to account for the enhanced effects along the axes in memory tasks, in matching tasks, and in spatial language tasks. Further, a number of findings suggest a distinction between the **vertical and horizontal axes**, including enhanced effects along the vertical axis in adult attention tasks (Logan, 1995), enhanced effects in location matching tasks (Landau, 1999, 2003), and earlier mastery of spatial terms along the vertical axis, relative to those along the horizontal axis (Landau, 1999, 2001; E. Clark, 1980). Within axial structure, there are separate effects of direction, revealed by studies of normal adults and children, and spatially impaired adults and children. The facts are consistent in suggesting that direction is represented separately from the axes themselves, and the evidence from spatial impairment suggests that direction may be a more fragile component of spatial representation than axial structure.

The theme of this volume is the representation of direction in spatial cognition and spatial language. I therefore will conclude by offering some observations on some of the similarities and differences between these two kinds of representations. To begin with, it should be said that much of the discussion on the relative usefulness of the two systems of representation could be clarified by a more precise characterization. In the literature, there appears to be considerable variability in what people mean by reference system, frame of reference, coordinate system, and vector representation. Significant progress awaits clear, precise, and consistent use of each of these terms.

The axial system I have described refers to a set of three orthogonal axes whose origin is centered on some Reference object or location (which may be determined by **attention**, see McCloskey and Rapp, 2000). Within each axis, we can additionally specify direction. An axial system need not represent distance, although it may. Furthermore, direction may be specified categorically—as in directions such as 'positive' or 'negative', or it may further specify metric direction, e.g. *47 degrees from X*. Vector systems (such as the one described by O'Keefe, 1996; Ch. 4) also specify an origin, and are metrically specified combinations of direction and distance.

To account for spatial language, it would appear that we require an origin, axes, and direction. Although it is possible for language to encode direction only (without any use of an axial system, as in *X ran away from Y*), many of the basic spatial terms appear to engage axial representations, but do so *without* specifying *metric* distance or direction. *Distance* is not required because spatial terms are categorical—that is, *X is above Y* whether it is 3 inches or 3 miles from Y. This may be modulated by object type, e.g. it is somewhat odd to say *the picture is 3 miles above the table* whereas it is fine to say *the airplane is 3 miles above the building*. The former is not semantically anomalous; it is just odd, due to properties of pictures and tables. However, this variation is object-specific, and need not be specifically represented in the meaning of the term *above*. **Metric direction** is not required because, again, the terms are categorical. Although the range of acceptable locations for *above* can be roughly defined in terms of angular distance from the origin, exact direction is not required except in order to determine whether it fits within this relatively large region of acceptability.

In contrast to the nice fit between spatial language and axial representations, it is hard to see how one can account for preferred usage of spatial terms by a vector representation. Of course, this would be possible by adding to the vector representation a specification of 'preferred' directions. Although this is possible, it is naturally accounted for by using axial structure as the basic representational kind. Similarly, although vector representations preserve metric distance, this does not appear to be required to capture spatial term usage.

What about non-linguistic tasks? Given the enormous variety of non-linguistic domains, it would seem that certain kinds of tasks might be better captured by a vector representation, whereas other kinds of tasks might be better captured by an axial representation plus direction. For example, category effects on memory for location have been documented by Huttenlocher and colleagues (Huttenlocher, Hedges, and Duncan, 1991; Crawford, Regier, and Huttenlocher, 2000). In these studies, people are asked to report the location of a dot in a display from memory. Their reports reflect axial organization, because they spontaneously impose both horizontal and vertical axes on the circle that holds the target dot. However, they also tend to displace the dots into regions whose centers migrate *away from* principal axes of the Reference object, and are located instead in the centers of the four quadrants formed by the horizontal and vertical axes. Huttenlocher and colleagues have suggested that memory for location is guided by categorical representations that are not isomorphic with axial structure. That is, another, *different* structure is either added to, or replaces, the basic axial structure in some tasks. In this case, neither axial representation nor vector representations would seem adequately to capture people's responses. In other cases—such as the tasks I have described—axial structure appears to be required, in order to capture the enhanced effects of memory or matching along the extended axes of the Reference object. In yet other cases, a representation that preserves metric distance and direction—such as a vector representation—would seem most appropriate. For example, as O'Keefe suggests, navigation appears to be guided by representations that preserve metric direction and distance. As another example, the case reported by McCloskey and Rapp (1999) demonstrates locational errors that preserve precise metric distance and axis, but destroy direction (*up/down, left/right*).

To conclude, axial representations appear to provide a natural vehicle for capturing many of the facts of spatial language. These must be enriched by specifying direction, and there is considerable evidence that axes and their directions may be represented separately. In contrast to the case of spatial language, it seems likely that non-linguistic spatial cognition as a whole will require a combination of representational types that are closely tailored to the nature of the cognitive task in question. Although at present there is no clear reason to place axial and directional representations in a strict hierarchy relative to each other, future research should clarify the contexts in which one or the other dominates.

3

Vectors across Spatial Domains: From Place to Size, Orientation, Shape, and Parts

JOOST ZWARTS

Abstract

Most of the literature about spatial language is restricted to place terms and prepositions. However, the domain of place does certainly not exhaust the language of space. There are at least four other spatial domains: size, orientation, shape, and spatial parts. The purpose of this chapter is to show how these domains fit together in a model of space that is based on vectors. In this way two important uses of vectors are integrated, the 'place use' (proposed by O'Keefe 1996 and Zwarts 1997 for prepositional relations) and the 'axis use' (used in Marr's 1982 3-D models to encode size, orientation, and shape in object representations).

Introduction

Every language has terms that are used to talk about space. The most prominent and most intensively studied of these spatial terms are the prepositions, the kind of words that in some sense have specialized in space, describing where something is or happens or where it is moving:

(1) a. The milk is *in* the refrigerator.
    b. We met *behind* the statue.
    c. They moved *into* another apartment.
    d. I led him *across* the yard.

I gratefully acknowledge the comments that I received from the audience at the Workshop on Axes and Vectors in Language and Space (see Preface), and from the editors. Postal address: UiL OTS, Utrecht University, Trans 10, 3512 JK Utrecht, The Netherlands. E-mail address: Joost. Zwarts@let.uu.nl.

Prepositions are one kind of what I will call **place terms** ((1a) and (1b)) and **change of place terms** ((1c) and (1d)). Members from other parts of speech can also be (change of) place terms:[1]

(2) a. He lives in the *vicinity* of a small town. (place noun)
    b. How *far* do you live from town? (place adjective)
    c. They *entered* the bar. (change of place verb)
    d. The airplane went *down*. (change of place adverb)

Most of the literature about spatial language is restricted to place terms and prepositions.[2] However, the **domain of place** does certainly not exhaust the language of space. There are at least four other spatial domains:

- The **domain of size**, with size terms ((3a) and (3b)) and change of size terms ((3c) and (3d)):[3]

    (3) a. a *thin* hair
        b. the *height* of a tower
        c. to *shorten* a rope
        d. to *grow*

- The **domain of orientation**, with orientation terms ((4a) and (4b)) and change of orientation terms ((4c) and (4d)):

    (4) a. an *oblique* line
        b. to stand *upright*
        c. he kicked it *over*
        d. two *rotations*

- The **domain of shape**, with shape terms ((5a) and (5b)) and change of shape terms ((5c) and (5d)):

    (5) a. a *straight* stick
        b. a *circle*
        c. to *straighten*
        d. to *bend* a stick

- The **domain of spatial parts**, with nouns such as the following:

---

[1] A class of place nouns that will not be discussed here are those referring to places in an absolute way, such as *place, site, area, spot*, etc.

[2] But see e.g. Lang (1990); Levinson (1992*b*); Landau and Jackendoff (1993); Bierwisch (1996); Cienki (1998); van der Zee (1996, 1997); and Schmidtke *et al.* (Ch. 9) for recent discussions of spatial terminology besides place terms or prepositions.

[3] The size adjectives, such as *thin* and *high*, are usually called dimensional adjectives (Bierwisch and Lang 1989). See also Bierwisch (1967); Fillmore (1971); and Teller (1969).

(6)   top, backside, corner, centre, bump, notch

This variety of spatial terms (and there may be other domains not included in this brief overview) raises the following questions: How do these different classes of terms within the language of space fit together? Is the domain of space a collection of unrelated portions of vocabulary or is there an underlying system? If there is a system, then how can we best analyze that system?

O'Keefe (1996), Zwarts (1997), and Zwarts and Winter (2000) have developed a semantics for spatial prepositions in which the position of a **Figure** (also called **located object, theme**, or **trajector**) relative to a **Ground** (also **reference object, relatum**, or **landmark**) is represented by a **vector**. Vectors are also used in Marr's 3-D models (1982) to represent the axes of objects and parts of objects and in this way they help to encode properties of size, orientation, and shape in object representations. This is also linguistically relevant as Landau and Jackendoff (1993) and Jackendoff (1996b) show for some terms of size and spatial parts (the 'axial vocabulary'). The purpose of this chapter is to adduce linguistic motivation for the integration of these two uses of vectors, the '**place use**' and the '**axis use**', by showing similarities and connections between the domain of place on the one hand and the domains of size, orientation, shape, and spatial parts on the other hand. The claim is that the relation between Figure and Ground in place terms is very similar to the notion of axis that plays a role in the meaning of terms of size, orientation, shape, and parts and that place relations and object axes are both represented by means of vectors.

I will first explain the vector-based approach to place terms in section 1. In section 2 I will show how the **Vector Space model** can be extended from the place domain to the domains of size, orientation, and spatial parts and what advantages that extension offers. In section 3 I will do the same thing for **paths** (sequences of vectors that represent change), to enable us to understand the language of shape and the language of change of size and orientation. Section 4 gives some final thoughts on the somewhat more complicated area of change of shape.

1. Place Terms and Vector Semantics

A **place term** is an expression that is used to locate an object or event. Place terms can be **static** (7) or **dynamic** (8) and they can be prepositions, adjectives, nouns, adverbs, or verbs:[4]

---

[4] The examples in (7) and (8) are taken from Dahl (1977).

(7) a. His wife was *out of* bed now, standing *beside* him *near* the window. (preposition)
  b. the big airbase at *nearby* Mildenhall (adjective)
  c. He was *around*. (adverb)
  d. in the *vicinity* of the town (noun)

(8) a. He moved *through* the half-daylight *over* the yard *to* the shed. (preposition)
  b. Gordon Butcher bent his head *closer and closer* still. (adjective)
  c. She went *downstairs* to make breakfast. (adverb)
  d. They *reached* the forward point of it. (verb)

One reason for using vectors in the semantics of place terms comes from modified prepositional phrases (Zwarts, 1997):

(9) a. the plow, which lay *ten yards* behind the tractor
  b. the plow, which lay *very far* behind the tractor
  c. the plow, which lay *just* behind the tractor
  d. the plow, which lay *straight* behind the tractor

The important thing about these **modifiers** is that they specify the distance or direction in which the located object (the plow) can be found relative to the reference object (the tractor). As a result they map the spatial region corresponding to the PP *behind the tractor* to a sub-region. It is hard to imagine how this works when the region is a set of points or a 'blob' of space because all by themselves points or blobs do not have distances or directions. However, suppose that *behind the tractor* corresponds to a **set of vectors** pointing backwards from the tractor and that the plow is located at the end point of one of these, as indicated in Fig. 3.1 (where the tractor and the plow are represented point-like and the shaded area corresponds to the end points of the vectors).

Modifiers can then be interpreted as mapping this set to a subset by imposing additional constraints on the length (*ten yards*, *very far*, and *just*) or orien-

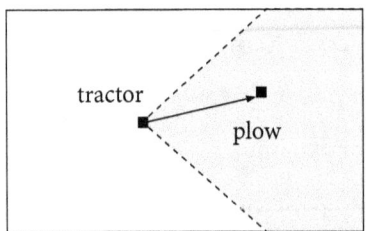

FIG. 3.1 *The plow behind the tractor.*

tation (*straight*) of the vectors. Furthermore, each preposition maps to a particular kind of region. *Near the tractor* denotes all those vectors that have a short length; *next to the tractor* corresponds to the vectors pointing sideways; *between the tractor and the barn* denotes the set of vectors pointing from the tractor towards the barn.

The kind of regions denoted by PPs in this theory are not unstructured sets. The algebra of vectors (with its operations of vector addition, scalar multiplication, inversion, and rotation) endows regions with a rich structure that allows for the definition of semantically significant classes of PP (see Zwarts, 1997; Zwarts and Winter, 2000, for more details).

A few remarks about **vectors** and **regions** are in order here. First, for the purposes of this chapter the precise formal or cognitive definition of vectors is of secondary importance. They can be taken as primitives, defined in terms of Cartesian or polar coordinates, as **pairs of points** (Gambarotto and Muller, Ch. 8), **bounded half-lines** (Schmidtke, Tschander, Eschenbach, and Habel, Ch. 9), or in terms of more complex **neurological patterns** (O'Keefe, Ch. 4). Of course, it does matter what a vector is, but not for the semantic phenomena under study here. Second, the vector representations diagrammed above are admittedly too simple in that they fail to account for the important fact that PP regions are really graded and vaguely bounded, as shown in Carlson, Regier, and Covey (Ch. 6) and the experiments cited there. They show how different factors operate on a vector representation to explain which parts of the region of a preposition receive higher acceptability ratings than others. For example, in an *above* region the points that are both close to the reference object and vertically aligned with it receive the highest ratings. Third, another simplification of the vector model presented here is the unwarranted exclusion of functional factors in the interpretation of prepositions, such as 'containment' for

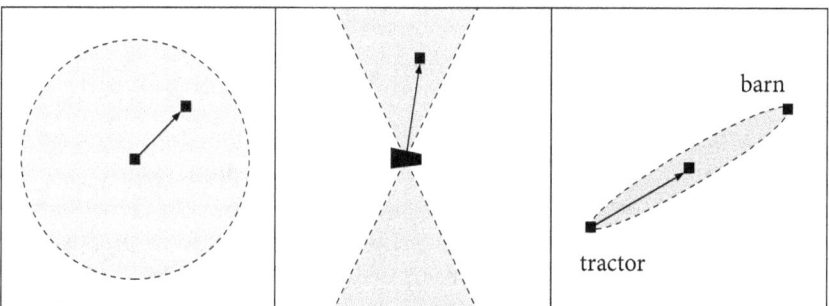

FIG. 3.2 *Near the tractor, next to the tractor, between the tractor and the barn.*

in, 'support' for *on* and 'protection' for *over* (see Coventry, Ch. 13). How to represent these functional notions in a mildly formal way and how to integrate them with the vector representation is unclear at present, although I personally feel that the concepts of **force dynamics** (Talmy, 1988) are crucial in understanding functions and relating them to spatial representations.

What about prepositions such as those in (10) that do not just locate an object but that help to express a **change of place** (10a), an **orientation** (10b), or the **distribution of an object** over a stretch of space (10c)?

(10) a. He went *into* the house.
      b. The sign points *away from* the village.
      c. The people were standing *along* the road.

A widespread assumption, which I will follow here, is that the underlying concept for these kind of expressions is a **path** (Jackendoff, 1983; Habel, 1989; Nam, 1995; Talmy, 1996, and many others). It seems intuitively a quite obvious move to analyze paths simply as vectors, as Helmantel (1998) does, but the problem is that paths often do not describe straight lines. For instance, *around the tree* can describe a path that takes the shape of a circle (among other shapes, see e.g. Schulze, 1991; Wunderlich, 1993; Taylor, 1995). Another option would then be to represent such a curved path as a **sequence of vectors connected head to tail**, as in Bohnemeyer (Ch. 5). See Fig. 3.3.

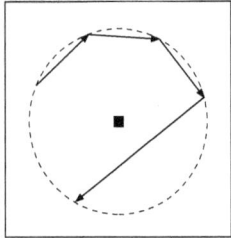

FIG. 3.3 Head–tail path.

Each vector of the path in Fig. 3.3 represents the move that the theme makes over a particular interval of time. The shorter the vectors of the path, the better the actual shape of a path is approximated. A path consisting of one vector could be seen as the simplest instance of this kind of representation.

Even though there are potential applications for these kind of paths (see sect. 3.4), I think that, as far as the interpretation of **directional prepositions** is concerned, there are good reasons to analyze paths first of all as **sequences of positions** relative to a reference object (Langacker, 1987; Jackendoff, 1996a), that is, in our framework, as sequences of vectors that all have one and the

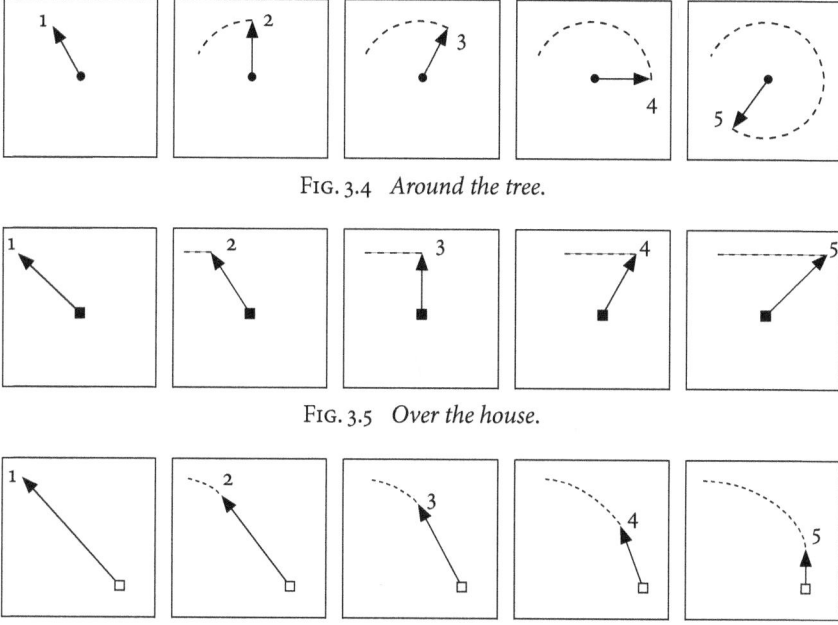

FIG. 3.4 *Around the tree.*

FIG. 3.5 *Over the house.*

FIG. 3.6 *Towards the city.*

same origin. For example, the PP *around the tree* would roughly correspond to a series of vectors like those in Fig. 3.4, *over the house* could refer to the sequence of vectors in Fig. 3.5, and *towards the city* corresponds to Fig. 3.6.

This representation captures in the most direct way the close relationship between position and motion and between locative prepositions and directional prepositions. It helps to understand more readily why a sequence of positions can be described by using a directional PP (Regier, 1995) and why entailments like these are valid:

(11)   a. The bird will fly *over* the house.
        b. The bird will be *above* the house.

Sentence (11a) entails sentence (11b) because the position *above the house* is just one of the positions that is part of the path over the house. There are also systematic mappings between locative and directional PPs:

(12)   a. The mouse went under the cupboard.
        b. The hotel is over the bridge.

In (12a) a locative PP *under the cupboard* is used to describe a path that ends under the cupboard. In (12b) a path *over the bridge* is used to specify a location as being the end point of the path. Representing paths as sequences of vector positions allows us to describe these mappings in a straightforward way.

Another reason for representing paths as sequences of vectors is that the length of the vectors of the path can sometimes be specified by a modifier, as in the following Dutch examples:

(13) a. De auto reed *vlak* langs de gracht.
 The car drove close along the canal.
 b. Het vliegtuig vloog *hoog* over de stad.
 The airplane flew high over the city.

Notice that the distance modifiers *vlak* 'close, right' and *hoog* 'high' do not specify a distance along the direction of the path, but a distance perpendicular to it.

All directional PPs receive the same type of path representation; there are not different kinds of representations for prepositions such as *to* and *from* and prepositions such as *around* and *along* (*pace* Bohnemeyer, Ch. 5). One important reason for having just one kind of path (apart from theoretical parsimony) is that different directional prepositions can be combined in a sentence to describe one path, as in our example (8a) above:

(14) He moved *through* the half-daylight *over* the yard *to* the shed.

There is only one path here, but it is described by three different directional PPs that specify different spatial positions of it.[5]

It is important that paths are atemporal entities, because they can be used in different ways, as we have already seen in (10): for motion (real or virtual), orientation, extension, and distribution. However, this notion of path can be extended to include time by relating the primarily non-temporal sequence to a sequence of moments of time, an operation that is necessary for describing motion. (See Jackendoff (1983, 1996a) and Talmy (1996) for more details and also sections 3 and 4 of this chapter. See furthermore Zwarts and Winter (2000) for more details on the vector analysis of paths.)

2. Place Vectors and Axis Vectors

2.1. Preliminaries

The claim of this chapter is that terms from the domains of place, size, orientation, shape, and spatial parts are all interpreted with respect to one and the

---

[5] Formally, each directional PP is interpreted as a set of paths and the asyndetic conjunction of PPs in this example as the intersection of the conjuncts.

same spatial structure, which has the form of a three-dimensional vector space V.[6] There is no need to assume different 'ontological categories', 'sorts', or 'primitives' for places, sizes, orientations, shapes, or spatial parts. Vectors are the only primitive spatial objects and each spatial domain constructs its meanings on the basis of vectors. Even if this claim will turn out to be too strong (which is quite likely), it still provides a fruitful perspective for studying the language of space, as I hope to show.

I assume that a vector v has two uses. It can represent either **an axis** of an object x, as in Marr (1982) or the **relative position** of an object x with respect to an object or point of reference y, as in O'Keefe (1996) and Zwarts (1997). This is illustrated in Figs. 3.7 and 3.8. I will use the term **place vector** for the vector in Fig. 3.7 and **axis vector** for the vector in Fig. 3.8. Even though the vectors are used in different ways, they come from one and the same vector space V. It is how they relate to objects that differs. I will represent this by means of two relations: **place** and **axis**.

An object x can relate to a place vector v in two different ways, depending on whether x functions as the Figure or the Ground of a spatial relation. Both of these functions are illustrated in Fig. 3.7: place(x,v) holds ('x is placed at vector v') as well as place(v,y) ('vector v is placed at y'). An object can be placed at more than one vector (as example (7a) shows) and there can be many vectors starting in one object, but there is always one unique vector for objects x and y (pointing from y to x). I will sometimes use place(x,v,y) to abbreviate the conjunction place(x,v) and place(v,y).

In Fig. 3.8 the relation axis(x,v) holds ('x has an axis v'). The assignment of axis vectors to objects is far more complicated than the assignment of place vectors and the treatment of axes can only be very rudimentary here. I will restrict the discussion to what are called **primary**, **major** or **generating axes** in

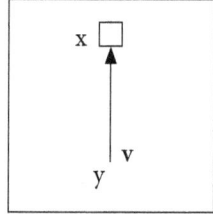
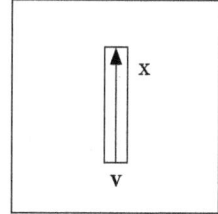

FIG. 3.7 Place vector.   FIG. 3.8 Axis vector.

---

[6] Actually, since many of the spatial terms will be interpreted on the basis of *located vectors* (vectors that have their starting point anchored in one particular spatial location) we need a vector space V_P for every point P, since each point is the origin of a vector space.

the literature (Marr 1982; Lang 1990; Jackendoff 1991; Levinson 1992b; Landau and Jackendoff 1993) and ignore the existence of **secondary, subsidiary,** or **orienting axes**. This restricts the domain of application to those objects that can be conceptualized as having only one axis, such as sticks, arrows, poles; and it excludes objects with more than one axis, such as windows, doors, and cylinders, as well as objects that do not seem to have axes, such as disks and balls. (Even with this restriction there still remains a lot to be said.) When an axis is assigned to an object it runs from one end to the other end and its direction is either arbitrarily chosen (as with a stick), or determined by the shape or function of the object (as with an arrow or spoon) (see Levinson, 1992b, for a discussion about this issue).

I will now show with concrete examples how this dual role of vectors as places and axes helps us to understand the underlying unity of the terminologies of place, size, orientation, and spatial parts. The treatment of terms referring to change of size or orientation, and the treatment of shape terms will be dealt with later, after I have explained the use of paths more fully.

2.2. Place and Size

As indicated already by H. Clark (1973) there is a close similarity between the place terms *close* and *far* on the one hand and *short* and *long* on the other hand, which is shown for instance by paraphrase relations such as those in (15):

(15) x is close to y ≈ the distance between x and y is short
      x is long ≈ one end of x is far from the other end

This similarity is brought out in the definitions for these expressions:

(16) x is close (to y): there is a v such that place(x,v,y) and $|v| < r$
      x is short: there is a v such that axis(x,v) and $|v| < r$
      x is far (from y): there is a v such that place(x,v,y) and $|v| > r$
      x is long: there is a v such that axis(x,v) and $|v| > r$

In these definitions | | gives the length of a vector v and r is some sort of context-dependent average for vector length. The y can be made explicit (*far from the house*) or left implicit (like the contextually given position of the speaker). Note that *close* and *short* both involve the same set of vectors, i.e. those vectors that are relatively small in length, pointing in all directions. The difference only consists in what those vectors are used to represent, place or axis. The fact that both place and size terms are based on vectors makes it easy to switch between place and size descriptions of the same situation, as in (15).

It has also been observed that the adjectives *high, low,* and *deep* (and the corresponding nouns *height* and *depth*) are ambiguous between a place and a size meaning (H. Clark, 1973; Lang, 1991; Taylor, 1995):

(17) a. a high window
b. a deep hole

The place reading of (17a) expresses that the window is at a high position relative to the Ground; the size reading makes a statement about its vertical dimension. Similarly, in (17b) either the hole is at a deep position, or it extends deep into the ground. What these terms have in common, both in their place and size use, is that the vectors involved have a **vertical orientation**, either upward (*high* and *low*) or downward (*deep*).[7] I will represent the upward direction UP as the set of vectors pointing upwards, and the downward direction as −UP (the inversion of all vectors in UP).[8] It is useful to have the concept of **projection of a vector on a direction**. As shown in Fig. 3.9, vector v has a unique projection v' on the vertical direction. This makes it possible to define a 'weaker' notion of upward and downward. A vector is upward if its projection on the UP direction is not the zero vector. This more inclusive region includes all vectors in the upper two quadrants, including the vectors of UP but excluding the horizontal ones. Similarly, a vector is downward if its projection on the −UP axis is not 0, which gives us the two lower quadrants. The definitions are then:[9]

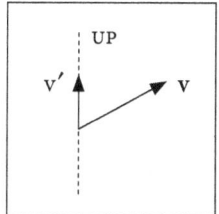

FIG. 3.9 Projection of a vector on the upward direction.

---

[7] This is a simplification of the meaning of *deep*. A hole that extends horizontally into a wall can also be described as deep, so it is more appropriate to define the direction of *deep* as inward from a surface.

[8] Possibly UP should be defined on the basis of a vector b that represents the canonical posture of the human body: UP = {sb: s > 0}, where sb is the vector that results when vector b is multiplied by a real number s. In this way the biological origin (H. Clark 1973) or embodiment (Lakoff, 1987) of the vertical dimension can be accounted for. −UP = {−v: v ∈ UP} = {sb: s < 0}.

[9] The definitions of the size adjectives *high*, *low*, and *deep* assume that the objects involved have a canonical upward position. The mast of a ship can be high, because its canonical position is upright, but arms are always long (or short), even if they are reaching up, because they do not have a canonical upward position.

(18) x is low (w.r.t. y): there is an upward v such that place(x,v,y) and |v| < r
x is low: there is an upward v such that axis(x,v) and |v| < r
x is high (w.r.t. y): there is an upward v such that place(x,v,y) and |v| > r
x is high: there is an upward v such that axis(x,v) and |v| > r
x is deep (w.r.t. y): there is a downward v such that place(x,v,y) and |v| > r
x is deep: there is a downward v such that axis(x,v) and |v| > r

(The implicit reference point y of the place terms is a contextually given ground level.) The ambiguity of these terms is restricted to the place vs. axis relations.

The vertical dimension with its two opposite directions UP and −UP is not limited to adjectives, as shown by the prepositions:

(19) x above y: there is an upward v such that place(x,v,y)
x below y: there is a downward v such that place(x,v,y)

If place and size terms are based on the same vector structure, it comes as no surprise that they share the same system of **measure phrases**.

(20) a. The stone was *twelve inches* deep.
b. The rope was *twelve inches* long.

There are no separate systems of unit nouns for distances on the one hand and for sizes on the other hand, but the same inches, meters, yards, etc. are used in both domains.[10] This shows clearly that place and size are basically the same domain, because different domains (like time, temperature, or weight) are usually characterized by their own system of units (e.g. seconds, degrees Celsius, kilograms). Measure phrases can be treated uniformly as additional restrictions on the vectors of place and size terms:

(21) x is *twelve inches* deep: there is a downward v such that place(x,v,y) and $|v|=12i$
x is *twelve inches* long: there is a v such that axis(x,v) *and* $|v|=12i$

---

[10] As a reviewer observed, this may be true for one-dimensional size, but not always for two-dimensional and three-dimensional size, where we have special unit nouns such as *acre* and *liter* (in addition to unit nouns based on the words *square* and *cubic*).

2.3. Place and Orientation

The major reason for building the semantics of place terms and orientation terms on a common foundation of vectors is the role that the **vertical and horizontal directions** play in both areas. Most orientation terms are interpreted relative to the framework of the vertical directions and the horizontal plane orthogonal to them. The framework sketched above for place and size terms extends naturally to orientation terms without much additional machinery. The simplest orientation terms state that an axis of x is pointing upward (22a), either downward or upward (22b), or that it is horizontal (22c):[11]

(22) a. x is standing, upright: there is a v such that axis(x,v) and v ∈ UP
 b. x is vertical, plumb: there is a v such that axis(x,v) and v ∈ VERT
 c. x is horizontal, level, lying etc.: there is a v such that axis(x,v) and v ∈ ⊥VERT

Other orientations can be can be described as falling between the vertical or the horizontal direction (23a) or as having a big or small angle with those directions (23b) and (23c):

(23) a. x is diagonal, inclined, sloping: there is a v such that axis(x,v) and v is between VERT and ⊥VERT
 b. x is steep, precipitous: there is a v such that axis(x,v) and the angle between v and ⊥VERT > r
 c. x is gentle, gradual: there is a v such that axis(x,v) and the angle between v and ⊥VERT < r

In other cases the axis of an object is reversed with respect to its canonical orientation vector c:

(24) x is reversed: axis(x,−c)
 x is upside down: axis(x,−c) and c ∈ UP

There are also orientation terms that take other directions R beside the vertical and horizontal as their 'reference direction', such as *straight* or *oblique*:[12]

(25) x is straight: there is a v such that axis(x,v) and v ∈ R
 x is oblique: there is v such that axis(x,v) and v is between R and ⊥R

---

[11] Like UP, VERT can also be defined on the basis of the 'body vector' b mentioned in n. 8: VERT = {sv: s is a real number}. The horizontal plane ⊥VERT is defined as the so-called 'orthogonal complement' to VERT, i.e. the set of vectors orthogonal to the vectors in VERT.

[12] Note that *straight* has an orientation meaning (with the opposite *oblique*) and a curvature meaning (with the opposite *curved*) that will be discussed in sect. 3.4.

The precise definitions are not so important. What is relevant here is that all the concepts used (axis, verticality, orthogonality, inversion, and even angle) are part of the same vector universe on which the domains of place and size are based.

Another indication of the intimate relation of place and orientation is that some adjectives of orientation can be used to modify prepositional phrases:

(26) a. *straight* above the window
     b. *diagonally* under the painting

This would be surprising if place and orientation were completely separate domains. In a vector, approach, however the modifiers *straight* and *diagonally* contribute one and the same vector condition either to axis vectors or to the place vectors of the prepositional phrases:

(27) x is *straight* above the window: there is a v such that v is upward and place(x,v,the-window) *and* $v \in$ UP
x is *diagonally* under the painting: there is a v such that v is downward and place(x,v,the-painting) *and v is between* –UP *and* ⊥–UP

Schmidtke *et al.* (Ch. 9) represent orientation and axes in terms of **half-lines**, which are like vectors except that they are unbounded in the direction in which they are pointing. As far as the domain of orientation is concerned their formalism covers the same ground as vector space semantics. However, because of their unbounded nature half-lines do not represent the length of object axes, even though this is relevant for dimensional adjectives. Either Schmidtke *et al.* need two different representations for axes, one for orientation aspects and one for size aspects, or they need to add a bounding point on the half-line, which is basically a way of defining vectors.

2.4. Place and Parts

That there is a close connection between **terms for parts** and place terms has often been observed, especially in relation to body parts (see for instance Heine, Claudi, and Hünnemeyer, 1991). Many languages use terms that are basically for parts of bodies or objects to talk about places adjacent to those parts. For instance, in English, the part noun *front* (as in *the front of the car*) forms the basis of the place term *in front of the car* and other part terms (such as *back*, *top*, and *side*) can also be used to denote places (*at the back of*, *on top of*, *to the side of*). In this case the intrinsic axes of objects are used to define regions outside those objects. Parts of objects can also be named on the basis of properties of the surrounding space. For example, a tree, an object without intrinsic horizontal axes, can be assigned a front and a back from the point of view of an observer and the top and bottom of an object can be determined by the direction of gravity.

It is intuitively clear that place and part terms share basic spatial concepts. For example, the opposition between *top* and *bottom* is of the same kind as the opposition between *above/over* on the one hand and *below/under* on the other hand. This is also illustrated by the fact that we can say that the top of an object is above the bottom and the bottom is below the top (if the object is oriented in its normal, 'canonical' way).

The general semantics of spatial parts requires a combination of the two vector relations, place, and axis. Figure 3.10 shows the schematic picture of a stick (y), with one end indicated by x.

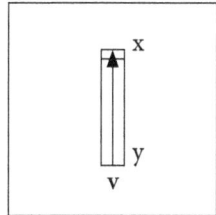

FIG. 3.10 A stick and one of its ends.

The stick y has an axis represented by the vector v and x is a part of y located at the end point of this vector. This can be represented as:

(28) x is an end of y: x is a part of y and there is a v such that axis(y,v) and place(x,v)

Many spatial parts will be parts that are located at a particular position relative to the axis of the whole. This is usually an extremity, but it can also be the middle:

(29) x is the middle of y: x is a part of y and there is a v such that axis(y,v) and place(x,$\frac{1}{2}$v)

The middle of the axis is derived by multiplying the axis vector by a scalar $\frac{1}{2}$, which yields a vector with half the length. The top and bottom of an object are defined as ends, but with an extra directional condition on the axis vector:

(30) x is the top of y: x is a part of y and there is an upward v such that axis(y,v) and place(x,v)
x is the bottom of y: x is a part of y and there is a downward v such that axis(y,v) and place(x,v)

These definitions are still an approximation because they do not distinguish between **intrinsic** tops and bottoms and **absolute or environmental** tops and bottoms (Jackendoff, 1996b; Levinson, 1996). A candle has an intrinsic top and bottom, because its canonical position is upright with certain defining features at each end. Even when a candle has fallen over, we can still talk about its top and bottom. A stick that happens to be in an upright position has a top and bottom in virtue of its position at that time with respect to the direction of gravity, but not an intrinsic top and bottom.

The front and back parts of an object can also be intrinsic (defined by its shape, function, or motion) or **relative or deictic** (defined by the position of an observer watching it). The following definitions will have to suffice here:

(31)   x is the front of y: x is a part of y and there is a vector v such that axis(y,v) and f(v) and place(x,v)
x is the back of y: x is a part of y and there is a vector v such that axis(y,v) and −f(v) and place(x,v)

The front and back of y are ends of y, but with an additional condition on the direction of the axis that captures the nature of the front–back axis (f for the front and −f for the back). Some possibilities: v can represent the direction of growth (Leyton, 1992), or the direction of motion, it can point to the side where the object is most commonly used or encountered (like the front of a television, E. Clark, 1973) or to an observer. For more details, see Herskovits (1986); Vandeloise (1991); van der Zee (1996); van der Zee and Eshuis (Ch. 12) and others.

How are places defined in terms of spatial parts? A possible way of doing this would be to take scalar multiplications of the axis vector. The range of numbers that are used for the multiplication determines whether the resulting place is external, in contact, or internal:

(32)   a. x is in front of y: there is a vector sv with s > 1 such that axis(y,v) and f(v) and place(x,sv)
b. x is on the front of y: there is a vector sv with s = 1 such that axis(y,v) and f(v) and place(x,sv)
c. x is in the front of y: there is a vector sv with $\frac{1}{2}$ < s < 1 such that axis(y,v) and f(v) and place(x,sv)

These three situations are illustrated in Figs. 3.11a–c.

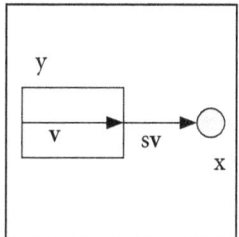 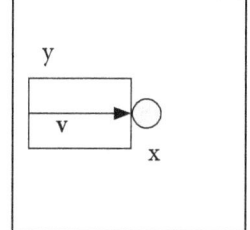 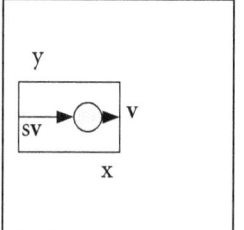

FIG. 3.11a. *X in front of y.*   FIG. 3.11b. *X is on the front of y.*   FIG. 3.11c. *X is in the front of y.*

3. Place Paths and Axis Paths

3.1. Preliminaries

Vectors alone are sufficient to represent the meaning of static terms of place, size, orientation, and parts. For the **representation of change or variation** (of place, size, and orientation) we need paths, as we saw in section 1 for the PP *around the tree*. We also need paths, as I will show, to provide a semantics for shape terminology.

A **path** can be defined as a sequence of vectors, or, more formally, a mapping from an interval [0,k] of real or natural numbers to vectors. Following Jackendoff (1983) and others I will take a path as a non-temporal entity, which means that the numbers used to define a path are part of an ordering device and do not represent moments of time. Figure 3.12 gives a simplified example of a path that describes a circle.[13]

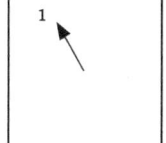 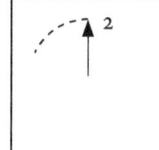 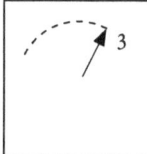 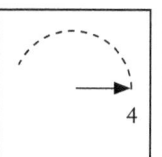 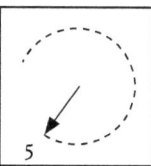

FIG. 3.12 A circular path.

[13] Depending on what is most convenient or clear, I will diagram paths either as a sequence of 'pictures' (e.g. Fig. 3.12) or as a constellation of numbered vectors in one picture (e.g. Fig. 3.15). This is just a matter of visual presentation and does not change the nature of paths as functions from numbers to vectors.

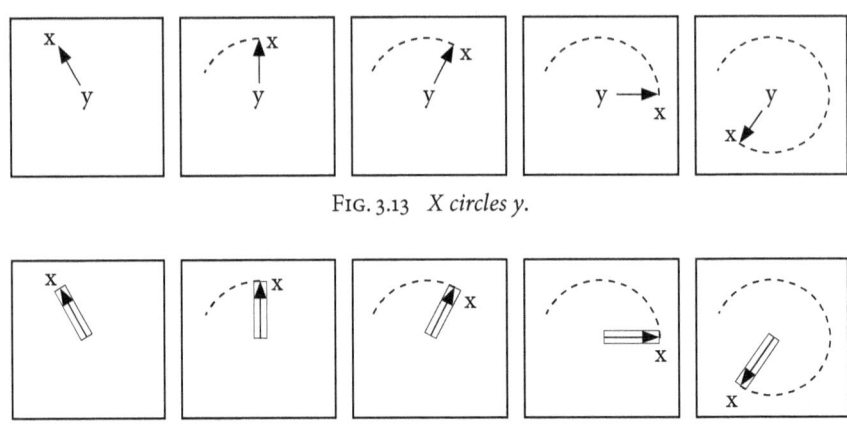

FIG. 3.13 *X circles y.*

FIG. 3.14 *X rotates.*

Like a vector, a path is an abstract spatial entity that can be used in different ways. The most obvious use is to describe the trajectory of an object in motion. The vectors function as place vectors, indicating the relative position of an object x moving through time relative to a point or object y, as illustrated in Fig. 3.13. I will call this a **change of place path**. Instead of functioning as place vectors the vectors of the path can also function as axis vectors, which represents that the axis of an object changes through time, a situation that is illustrated in Fig. 3.14 and that I will call a **change of axis path**.

I propose to interpret the two verbs *to circle* and *to rotate* in this way:

(33) to circle y {e: there is a p such that place-path(e,p,y) and p is circular}

to rotate {e: there is a p such that axis-path(e,p) and p is circular}

The verbs (and the verb phrases and sentences built around them) are interpreted as sets of events, which is what the 'e' stands for. An **event** can relate to a path in two ways: (1) by means of the relation place-path that relates an event to the path that that event describes through space (relative to y), what is sometimes called the spatial 'trace' of the event (Krifka, 1995; Link, 1998), (2) by means of a relation axis-path that associates an event to the path of a changing axis. Both functions are defined in terms of the more basic relations place and axis of section 2. If an event e has a particular path p as its path of motion, then for each moment of the event, its theme (the thing that is moving, the subject of *to circle*) is located at the corresponding vector of the path. And for change of axis, a path p represents the way an axis changes if and only if for each moment of the event the axis of the theme is identical to the corres-

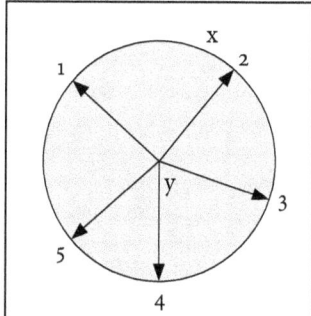 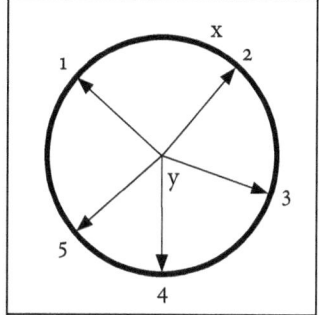

Fig. 3.15 *A round disk.*   Fig. 3.16 *A round ring.*

ponding vector of the path.[14] A path is circular if it is the result of rotating the initial vector p(0) of the path the full 360°.[15]

In addition to using a path to represent properties of motion or rotation events we can also use paths to represent shape properties of contours or linear objects. The path of Fig. 3.12 can be used to represent the contour of a disk or the circular axis of a ring, as illustrated in Figs. 3.15 and 3.16, respectively.

The adjective *round* can then be defined as follows:

(34)   round {x: there is a p such that place-path(x,p,y) and p is circular}

The same function place-path that was used for *to circle* is used here to relate an *object* x to a path p (relative to y, which is here the center of the circle). In this case there is not an association between temporal parts of the event to vectors on which to locate the theme, but there is an association between spatial parts of the object x and vectors on which those parts are located. When applied to the situation in Fig. 3.15 the parts of the contour of the disk are related to the vectors of the path. In Fig. 3.16, the parts of the ring are related to

---

[14] If we assume (following Krifka, 1992 and Jackendoff, 1996a, among others) that an event e corresponds to a temporal interval time(e), that the theme of the event is given by theme(e), and that the relations place and axis have a temporal parameter t, then place-path and axis-path are defined as follows:

place-path(e,p,y) iff for every t ∈ time(e) place$_t$(theme(e), p(μ(t)),y)
axis-path(e,p) iff for every t ∈ time(e) axis$_t$(theme(e), p(μ(t)))

with μ an isomorphism from the interval time(e) to the domain of the path.

[15] More precisely, the set of circular paths would be {p: ∃ρ ∀i p(i) = ρ(i)p(0)} where ρ is a function that maps every element i of the domain [0,k] of the path to a rotation ρ(i) which is then applied to the initial vector p(0). In the 'prototypical' instances of around, ρ is a continuous, monotone increasing function with ρ(0) the zero rotation and ρ(k)p(k) = p(0), i.e. the path goes all the way round once with a constant radius. For many terms based on circular paths (like *around, to surround*, etc.) we need a more flexible definition.

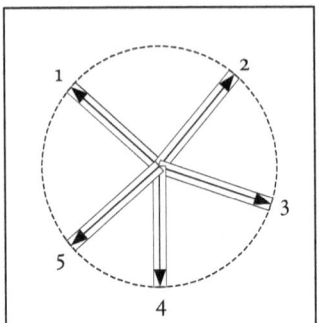

FIG. 3.17 *Spiral staircase.*

those vectors.[16] (Of course, the path describing a circle would need more than just the five vectors drawn here for the sake of the presentation.) I will call this kind of path a **shape path**.[17]

Can axis-path also apply to objects? What we would need in this case is an object that can be divided into a sequence of parts each of which has an axis and these axes show a variation in direction or length. For circular paths, one example might be a spiral staircase, where the stairs wind round a central pillar. Each stair has a vector that points from this center and (when seen from above) the vectors together form a circular path, as illustrated in Fig. 3.17.

This situation can be described in the following way:

(35) there is a p such that axis-path(x,p) and p is circular

An additional complication of the spiral staircase is its vertical dimension. A more complete representation involves two paths: the circular path defined in (35) and a path of vertical vectors representing the heights of the stairs either from the ground up or from the top down. Each step of the staircase is then characterized by two vectors: a place vector for its height and an axis vector for its length and orientation.

Recall from section 1 that there are different ways of representing paths in terms of vectors. Instead of describing a path as the end points of vectors

---

[16] For place-path to apply to objects we need a function called extent that maps contours and extended objects to the ordered set of their paths:

place-path(x,p,y) iff for every $x' \in$ extent(x) place(x',p(μ (x')),y)

Again there is an isomorphism μ that maps the parts of x onto the domain of the path (see also Jackendoff, 1996a).

[17] Describing the shape of a three-dimensional object, such as an apple, by means of a path is a bit more complicated. I assume that a three-dimensional object can be called 'round' because each of its cross-sections can be called 'round' as defined in (34).

coming from one central origin, we can also construct a path by putting vectors head to tail. For 'circular' expressions such as *round, around, to circle, to rotate*, and *spiral* a path with a central origin is required, but for other terms, a 'head–tail path' seems more appropriate. Which representation is appropriate depends on the lexical semantics of the expression under consideration, and maybe on other factors. We might even be able to switch between the two modes of representation.[18]

With this background concerning paths let us turn to the various spatial domains.

3.2. Paths in Place and Size

Size cannot always be represented on the basis of vectors, for the simple reason that many things that have size are not straight, but curved in various ways. The dimensional adjectives *long* and *short* apply also to ropes and rivers just as much as to sticks and streets and the size nouns *circumference* and *perimeter* measure something that excludes any idea of straightness. We saw that the contour or axis of a curved object can be represented by a shape path, which is illustrated in Figs. 3.18 and 3.19 for a part of a river. Although any point could be taken as the origin of the vectors describing the centered path, it seems most natural to take one end of the river as the origin, as in Fig. 3.18. Alterna-

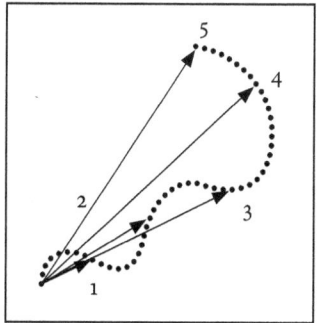
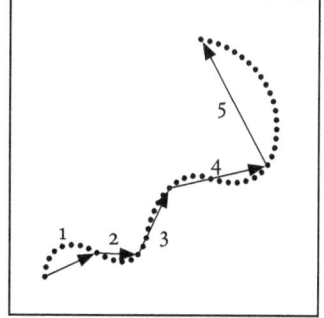

FIG. 3.18 Centered path of a river.　　FIG. 3.19 Head–tail path of a river.

---

[18] Any 'centered' path p can be mapped to a head–tail path p' in a systematic way, by connecting the end points of p: if p is a centered path over a domain of integers [0,k], then the corresponding head–tail path p' with domain [0,k−1] assigns to every i the vector that connects the end points of p(i) and p(i+1). Conversely, given a central point O we can turn a head–tail path p with domain [0,k] into a centered path p' with domain [0,k+1]: for every i ∈ [0,k], p'(i) is the vector that connects O and the starting point of p(i) and p'(k+1) is the vector that connects O with the end point of p'(k).

tively, we can describe the shape of the river by means of a head–tail path, as in Fig. 3.19. In both cases we need to choose an appropriate scale or grain size of our representation by fixing the number of vectors we use to represent the shape. Since *long* and *short* can apply to straight and curved objects, I assume that their definitions come in two varieties, for axis vectors in (36a), and for shape paths in (36b):

(36) a. x is long: there is a v such that axis(x,v) and $|v| > r$
x is short: there is a v such that axis(x,v) and $|v| < r$
b. x is long: there is a p such that place-path(x,p) and $|p| > r$
x is short: there is a p such that place-path(x,p) and $|p| < r$

The length function | | would have to be defined for paths in an appropriate way.

There are cases where a size is described in terms of the directional prepositions *around* and *across*:

(37) a. The crater was two miles round.
(i.e. The circumference of the crater was two miles.)
b. The crater was two miles across.
(i.e. The diameter of the crater was two miles.)

The circumference of the crater, a size concept, is described in terms of the length of a movement around it and its diameter in terms of the length of a movement across. Notice also that the nouns *circumference*, *perimeter* and the Dutch *omtrek* are all based on prepositions meaning 'around': *circum* in Latin, *peri* in Greek, *om* in Dutch. The word *diameter* and the corresponding Dutch word *doorsnede* are based on prepositions meaning 'through, across', *dia* in Greek and *door* in Dutch. This is not surprising if the same kind of paths are used in motion and shape description: *circumference* and *round* are both based on circular paths, and *diameter* and *across* on paths that go from one side of an object to the other side. The measure phrases in (37) specify the lengths of these paths.

Paths can be relevant for size in another way. When the size of an object changes, it is the axis vector that becomes longer (in the case of *to grow*, *to extend*) or shorter (in the case of *to shrink*, *to reduce*). The interesting thing is that the *direction* of change of size can be specified by means of a phrase that is also used for motion:

(38) a. The town is extending *to the east*.
b. John is growing *up*.

I interpret these sentences as sets of events:

(39) a. {e: there is a p such that axis-path(e,p) and theme(e,the-town) and the length of the vectors of p is increasing over its domain *and the vectors of p are pointing to the east*}
b. {e: there is a p such that axis-path(e,p) and theme(e,john) and the length of the vectors of p is increasing over its domain *and the vectors of p are pointing upward*}

The italicized expressions in (38) correspond to the italicized parts in (39).

3.3. Paths in Place and Orientation

The terminology for change of orientation (i.e. rotation) overlaps with terms in the place domain in several ways. Words such as *around* and *turn* can be used to describe the rotation of an object that stays in one place (40a) but they can also be used to describe the path of a moving object (40b) and (40c):

(40) a. He turned around to see what had happened.
b. The car turned into a small street.
c. She ran around.

The ambiguity of the verb *turn* and the adverb *around* depends on whether the underlying circular path is used as a change of *axis* path (in (40a)) or as a change of *place* path (in (40b) and (40c)).[19]

On the other hand, the adverbs *up* and *down* are not only used for motion up and down, as in (41a) and (41b), but also for rotation (from horizontal to vertical or vice versa), as in (41c) and (41d):

(41) a. The ball fell down.
b. The bird flew up.
c. The tree fell down.
d. The mast was pulled up.

The tree in (41c) and the mast in (41d) describe paths that are one-quarter of a circular path, but what is more important here is the projection of their axis on the vertical directions. The projection of the tree's axis on the upward direction is decreasing, the projection of the mast's axis is increasing.

It has been observed (Jackendoff, 1983; Talmy, 1996) that verbs of orientation can be modified by path PPs (42a) and (42b) and also that the orientation of

---

[19] The notion of circular path needs to be extended for these cases. The paths of these three sentences can take the shape of a semi-circle and in (40c) the path is most likely rotating in a more 'random' way in combination with strong variation of the length of the vectors.

an object in relation to another object can be described by means of path PPs (42c) and (42d):

(42) a. The stick is aligned towards the east.
b. The arrow is pointing across the lake.
c. The tree is lying along the railway.
d. There is a scratch diagonally across the door.

This is another instance of terms of orientation and terms of place interacting. The simplest spatial representation of the straight objects in (42) is a single vector. However, in these examples the orientation is described by means of directional PPs, which I assume to be interpreted in terms of paths (sequences of vectors). Hence, there is a mismatch between the subject (one vector) and the predicate (sequence of vectors). In order to resolve this mismatch I assume that axes of straight objects can be represented in two ways: by means of single vectors, but also by means of straight shape paths (i.e. all the vectors point in the same direction). This captures the idea of Talmy (1996) that we can ascribe 'fictive motion' to elongated stationary objects, for example by our gaze moving from one end to the other, and it makes it possible to understand how the PPs can apply to these non-motion cases:[20]

(43) a. there is a p such that place-path(the-stick,p,y) and p is towards the east
b. there is a p such that place-path(the-arrow,p,y) and there is an extension p' of p such that p' is across the lake
c. there is a p such that place-path(the-tree,p,y) and p is along the railway
d. there is a p such that place-path(a-scratch,p,y) and p is diagonal and p is across the door

The y in these definitions is one end of the object, chosen as the starting point of the shape path. Whether the directional PPs in (42) are used in this kind of sentences or in real motion sentences makes no difference for their path denotation. In motion sentences the same paths are not related to an object but to an event:

---

[20] In these definitions and elsewhere in the chapter I leave PPs such as *towards the east* undefined because the focus is not on the lexical semantics of directional prepositions. See Zwarts and Winter (2000) for a proposal on how to define various directional prepositions. Note that *towards the east* is a bit special because there is not a specific point called *the east* that we can take as the origin of the vectors of the path, like we did with *towards the city*. In some sense, *towards the east* expresses a direction, rather than a destination, and it seems more reasonable to represent it as a path of vectors of increasing length pointing east. How to unify the semantics of *towards* in such a way that it will cover both uses is a topic for future research.

(44) a. The bus drove towards the east.
b. there is an e and a p such that place-path(e,p,y) and p is towards the east

Both in (42a) and in (44a) the denotation of the PP *towards the east* is a sequence of vectors all pointing to the east and increasing in length.

3.4. Paths in Place and Shape

In the domain of shape we see more examples of ambiguities and interactions. Several basic shape terms can also be used with place meaning. The shape adjective *round* forms the basis of the adverb/preposition *(a)round* and the noun *circle* for the verb *to circle*. All these terms are based on circular paths that are either associated to events (*to circle around*) or to objects (*a round circle*).

Path PPs can also be used to modify shape terms:

(45) a. The coast curves *outward into the sea.*
b. The Chinese wall zigzags *through the hills.*

The coast and the wall are not moving along the paths denoted by the italicized PPs, of course. What happens is that there is an underlying path p in each sentence associated with the coast and the wall respectively that is further restricted by the verb (*to curve, to zigzag*) and by the PP:

(46) a. there is a p such that place-path(the-coast,p,y) and p is curving and the vectors of p are pointing outward and into the sea
b. there is a p such that place-path(the-Chinese-wall,p,y) and p is zigzagging and the vectors of p are pointing to positions in the hills

This is illustrated in Figs. 3.20 and 3.21. Even though it might not be immediately obvious how to define curving and zigzagging of paths and 'outward',

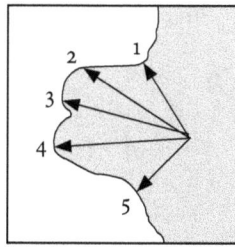
Fig. 3.20 The coast.
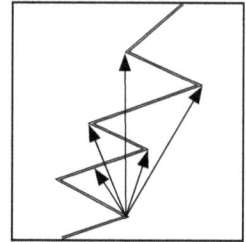
Fig. 3.21 The Chinese wall.

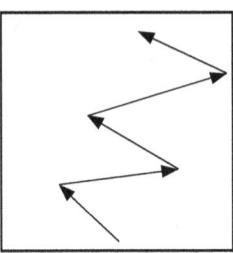

Fig. 3.22 A zigzag path.

'into', and 'through', still the rough definitions and figures demonstrate that an abstract notion of path, generalizing over the domains of place and shape help us to understand the semantic structure of sentences such as those in (45). Note that for zigzagging paths a head–tail path might be more appropriate to capture the 'zigzagging' shape than a centered path (Fig. 3.22).

In the representation a head–tail path p is zigzagging iff every two subsequent vectors make an acute angle. Whether we choose the representation in Fig. 3.21 or Fig. 3.22, the important thing is that zigzag paths are made up of component vectors that represent the individual back and forth motions characteristic for zigzagging. Modifiers can specify the direction of these individual vectors, as in this example (*in and out, back and forth, across the wake of the motorboat*):

(47) In the slalom event, the towing boat speeds straight through a field of anchored buoys while the contestant, riding on a single ski or a kneeboard, pursues a zigzag course in and out of the buoys, swinging back and forth across the wake of the motorboat.[21]

3.5. Other Interactions

Until now we have seen the interactions of the domain of place with various other domains, but there are also connections among the domains of size, orientation, shape, and spatial parts involving paths:

(48) a. The tree is straight.
b. De boom groeit scheef/krom.
   The tree grows askew/curved.
c. the circumference of the disk

---

[21] Taken from 'Waterskiing', Microsoft® Encarta® 98 Encyclopedia. ©1993–7 Microsoft Corporation. All rights reserved.

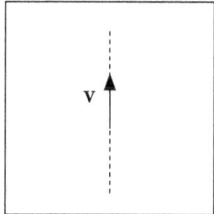

FIG. 3.23 Orientation *straight*.

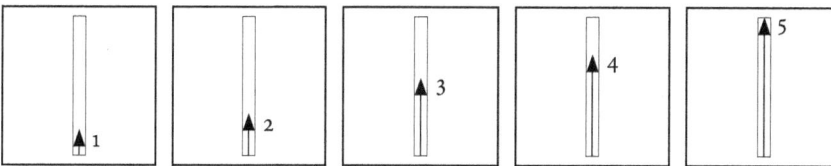

FIG. 3.24 Shape *straight*.

The word *straight* in (48a) is ambiguous between an orientation and a shape meaning. It can be the opposite of *askew/oblique* or it can be the opposite of *curved*. These two meanings are closely related. The orientation meaning applies basically to one vector, saying that that vector makes a zero angle with a reference direction, as shown in Fig. 3.23, the shape meaning applies to a path of vectors, roughly saying that each vector p(i) of the path points in the same direction, as shown in Fig. 3.24. In other words, some basic shape concepts, such as *straight* and *curved*, can be defined in terms of the relative orientation of the vectors making up the shape paths of the objects to which these terms apply.

In the Dutch example (48b), the change of size verb *groeien* 'to grow' can be modified by an orientation term *scheef* 'askew' or a shape term *krom* 'curved'. The central part of the sentence states that there is a path p associated with the tree and the vectors of this path are increasing in length (growing). The adjective *scheef* adds the information that all these vectors are making a particular angle with the vertical direction. The adjective *krom* says that the vectors of the path make an increasing angle with the vertical direction. Figures 3.25 and 3.26 show the two processes of growing.

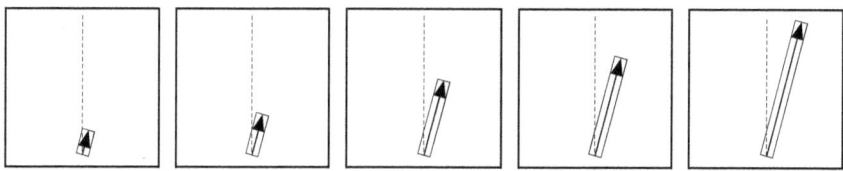

FIG. 3.25 'Growing askew'.

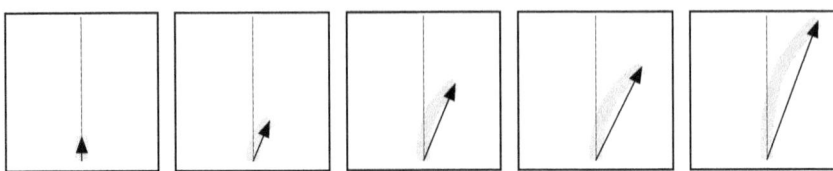

Fig. 3.26 'Growing curved'.

Finally, (48c) shows an ambiguity between a part meaning and a size meaning. The circumference of the disk can be either a part of the disk, i.e. the rim, or it can be the length of that rim. Underlying both of these meanings is the circular path that was discussed earlier. The size meaning of *circumference* focuses on the length of this path, the part meaning takes the whole of all the parts of the disk located on this path.

3.6. Change of Shape

The basic entity of the model I am describing is the vector. Vectors can be used to capture the meaning of static terms of place, size, orientation, and spatial parts, in so far as these terms are based on a relative position or axis that can be conceived as a straight line. Paths are constructed out of vectors and are used to represent change of place, size, and orientation, as well as sizes, shapes, and parts that cannot be represented by means of a vector. However, a more complicated object seems to be necessary to represent **change of shape**. Suppose we want to represent the meaning of sentence (49):

(49) The man bent the iron bar.

At the beginning of the bending event the shape of the bar can be represented by means of a straight path, as explained a few paragraphs back. At the end of the event we have a bar with a curved path. Each stage of the event is associated with a path representing the axis of the bar, or, maybe, the event as a whole is associated with a sequence of paths, a 'hyper path'.

One thing that should be accounted for is the possibility of modifying the direction of shape change by means of directional PPs:

(50) The bar was bent to the left.

One way to tackle this example is to say that the shape path at the end of the event is a path that is in the denotation of the PP *to the left*. This is shown in Fig. 3.27 by means of a sequence of paths.

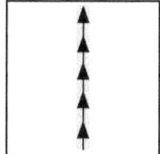 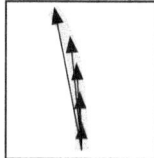 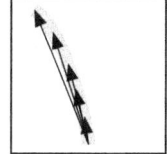 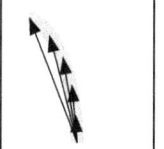 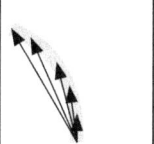

FIG. 3.27 To bend to the left.

I also mention in this respect, without providing an analysis, that motion and static shape can also be described in terms of curving or bending:

(51) a. The jet fighter curved back to its formation.
      b. The road bends to the left.

Even though the analysis is not obvious, these examples again illustrate the basic claim of this chapter, that various spatial domains are intimately related.

Conclusion

One way of evaluating the explanatory value of a theoretical notion is to extend its domain of application as far as possible. If vectors are really a fundamental concept in spatial language and spatial cognition, then we expect them to be relevant to the analysis of more phenomena than, let's say, modified prepositional phrases. I have tried to show that there are in fact good linguistic reasons to treat vectors as the backbone of spatial representation across the spatial domains of place, size, orientation, shape, and spatial parts. Even though some of the analyzes are still a bit sketchy the general direction of research seems promising.

One critical conclusion that one might draw from the wide range of applications for vectors in spatial language is that vector space semantics is really too powerful. There seems to be no limit on what you can do with it. My conclusion, however, is that the comprehensiveness of vector semantics does not so much demonstrate the power of the formalism, but rather the conceptual economy of the spatial domain. A vector is really a very simple thing: a combination of a length and a direction. If so much of our spatial language can be analyzed in terms of vectors, then my conclusion would be that our conceptualization of space is almost exclusively based on these two notions.

On the other hand, as shown in Zwarts (1997) and Zwarts and Winter (2000), this does not mean that any mathematical construction based on vectors (any vector set or vector path) is a possible meaning for a spatial term. The denotations of prepositional phrases are subject to strong algebraic constraints and I would expect the same for terms of size, orientation, and shape. One of the aims for future research is to find the algebraic properties that are characteristic of meanings in the spatial domain as a whole as well as for meanings in specific sub-domains. Obviously, such a program only makes sense when a model is used in which the right algebraic properties can be formulated in the first place.

4

Vector Grammar, Places, and the Functional Role of the Spatial Prepositions in English

JOHN O'KEEFE

Abstract

In a previous work, I presented a theory of the role of the locative prepositions in English which suggested that their primary function was to identify a set of vectors that located regions in a Narrative Map, a device for storing narratives and other linguistic representations. In this chapter, I present an updated version of this idea. Here I draw upon recent computational ideas generated by our efforts to model the firing fields of the spatially coded place cells in the rat hippocampus. The chapter attempts to show how different prepositions specify different aspects of the location vectors and of the spatial extents of the fields in different directions.

Introduction

I wish to update my proposal (O'Keefe, 1996) that the neural basis for language might be a structure termed a **Narrative Map** in which syntactic relationships are specified by places, the names of the objects that they contain, and the vectors connecting them. This is a proposal that Lynn Nadel and I first mooted in our book *The Hippocampus as a Cognitive Map* (1978) and which is motivated by neural as well as linguistic considerations. We have proposed that the **hippocampus**, a cortical area in the mammalian forebrain, is involved in the construction of an allocentric spatial representation of the environment, what

---

I would like to thank my colleagues Neil Burgess and Tom Harley for discussions and earlier collaborations on the computational modeling of place cells and Boundary Vector cells. That research was supported by an MRC program grant.

Tolman (1948) called a '**Cognitive Map**'. In animals, this map is purely or primarily spatial: the hippocampal formation provides the animal with a representation centered on the environment which enables it to locate itself and objects of interest within that environment. In humans however, the hippocampus is necessary for the storage and recall of linguistic and episodic memories as well as for spatial memories and so it was necessary to extend the basic theory to account for the human data. We assume that the right human hippocampal formation continues to have a primarily spatial function, operating in much the same way as the rat hippocampus. The left hippocampus, on the other hand, has been modified in two ways to transform it into a linguistic/episodic memory system. Firstly, this spatial structure has acquired an extra dimension enabling it to incorporate a temporal sense into the basic map to account for the ability of humans to process and store spatio-temporal information. We argue that this framework provides the basis for **episodic memory**, which is the ability to recall personally experienced events set within their original spatio-temporal context. Secondly, we argue that the primary input to the left hippocampal formation consists of information about **linguistic entities** rather than about physical objects referred to the external physical world. For example, damage to the left mesial temporal lobe usually results in impairment in the memory for linguistic material and in particular words and narratives (see e.g. Frisk and Milner, 1990). I will briefly outline our current understanding of the neural elements which contribute to the Cognitive Map and then I will sketch our recent computational model of the formation of the place representations that form the core of the Cognitive Map and show how some aspects of this model relate to the prepositions in English.

1. Cognitive Map Theory

The Cognitive Map consists of a set of place representations and their spatial relationship to each other, objects are located by their relationship to places in the map and only indirectly by their spatial relationships to each other. Spatial relationships are specified in terms of three variables: places, directions, and distances. **Places** are patches of an environment that can vary in size and shape depending on the size of the environment and the distribution of features in that environment. They are located in two ways: first, in terms of the spatial relations amongst the invariant features of the environment and second and independently, by their direction and distance from other places. The place code is carried by the pattern of firing of the place cells in the cortical region called the hippocampus. **Directions** are specified as a set of parallel, infinitely long vectors. As with places, these can be identified in one of several ways: either as the local gradient of a universal signal such as gravity, geomagnetism, or olfactory currents, or as the vector originating at a place or object and passing through another place or object (or passing through two places), or as having a specified angle to a previously identified direction (e.g. through updating the current direction on the basis of angular head movements). For every direction there is an opposite direction which can be marked by the negative of that vector. The direction code is carried by the pattern of firing of the head direction cells, which are located in several brain regions but most notably in the dorsal presubiculum (see e.g. Taube *et al.*, 1990), a cortical region that neighbors on the hippocampus and is anatomically connected with it. **Distances** between objects or places are given by a metric, which can be derived in one of two ways. Ultimately distances are based on a signal that converts information about the organism's movements in an environment into the speed with which those movements translate the animal through the environment. The integral of this speed signal is combined with introceptive (e.g. vestibular) and/or environmental (e.g. directional) information to calculate a distance between places. The changes in the relative locations of objects as a result of movements in the environment can also be used to compute distances.

A **path** is defined as an ordered sequence of places and the direction vectors between them. Paths can be identified by their end places or, in humans, by a distinct name. Conversely, places along the path can be identified and associated with the path. A path may be marked by a continuous feature such as an odor trail or a road, but need not be.

Within this spatial framework, translations of position in an environment are specified as **translation vectors** whose tail begins at the origin of movement and whose head ends at the destination. Vector addition and subtraction allow journeys with one or more subgoals to be represented and integrated. Furthermore, on a journey with more than one destination the optimal or minimal path can be calculated. In recent computational work, our group has suggested that the activity patterns of the cells which encode the place representation might be constructed on the basis of inputs that identify a location on the basis of its direction and distance from large environmental features such as a wall (Hartley *et al.*, 2000). Here I will pursue a variant of this idea and see whether the spatial prepositions can be identified with different aspects of this computational model.

I have previously argued that an understanding of sentences about physical space might provide an important insight into the deep semantic structure of language in general. There is a long tradition in linguistics, revived within Case Grammar theory, which postulates that the deep semantic structure of language is intrinsically spatial and that other, non-spatial, prepositions are in some way parasitical on these prototypical formulae, perhaps by means of metaphorical extension of their core spatial meanings. This is the contention of a group of linguists called **locationists** or **localists** (Anderson, 1971; Bennett, 1975) (see Cook, 1989, for a review). In an uninflected language such as English many of the spatial relations described in spatial sentences are conveyed by the prepositions. A description of the representations set up by the spatial prepositions might provide the basis for a more general linguistics. Nadel and I opined that the origin of language might have been the need to transmit information about the spatial layout of an area from one person to another (O'Keefe and Nadel 1978: 401 n.), perhaps as adjuncts to simple maps used to convey the location of food items or dangers to other family members. Originally the linguistic content of these prototypical **Semantic Maps** might be rather simple and impoverished. Different sounds might stand for different objects in the map and might serve the additional function of acting as an encrypting device. Over time the pictorial aspects of the structure of the map might be systematically replaced by prepositional and other spatial semantic elements. This increase in syntactic vocabulary would eventually obviate the need for the externalized map entirely but the neural substrate would retain the underlying map-like structure of the original.

In O'Keefe, 1996, I set out the basic framework of Vector Grammar and showed how it accounted for many of the spatial meanings of the spatial prepositions. My thesis was that the primary role of the prepositions was to provide the spatial relationships among a set of places and objects, and to specify movements and transformations in these relationships over time. These spatial relationships and their modifications were viewed as represented by vectors. A similar idea was proposed by Zwarts (1997) around the same time.

In **Vector Grammar Theory**, the location of an entity within this notation is by a vector, which consists of a direction and a distance from a known location. Some of the work of the locative prepositions involves the identification of aspects of these variables. In some cases (e.g. with vertical prepositions such as *below*), the direction is given by an environmental signal such as the force of gravity. In most cases, however, it needs to be calculated from the spatial relationships between two or more objects or places. In these latter cases the prepositions specify the origin and termination (or the tail and the head) of the vector or a point along the vector. In contrast, distances are less well specified; in most cases the metric is an *interval* one. This reflects the fact that many prepositions describe relationships that are transitive, linearly scalable, and insensitive to absolute location. If A is *above* B and B is *above* C, then it follows that A is *above* C. Doubling the distance between A and B, and between

B and C does not change any of these relationships. Nor does shifting all three entities a constant distance in the same direction. One of the roles of the preposition *for* is to supply the necessary metric information. When this is available, the relationships described by the spatial prepositions can attain the level of a *ratio* scale (for further discussion of these ideas, see O'Keefe, 1996). The 3-D space coded by the locative prepositions is a mixed polar-rectilinear one.

In this chapter I shall explore a modified version of the same **vector-based approach**. Many of the ideas it contains are derived from extensions of the recent work of our group in which we have proposed computational models for the place fields and their cortical inputs. In this work we have suggested that places can be modeled as the sum of two or more gaussians. The center of each **gaussian** is located by its distance in a particular direction from an environmental feature, i.e. a vector. Several of these gaussians are combined in the hippocampal formation to form a place representation. It is my contention that some of the prepositions code for aspects of single directional vector gaussian inputs to the hippocampus whereas others seem to require more complex representations anchored in several directions, perhaps even ones similar to the place cells themselves. I assume (following the locationists, see above) that the prepositions in English have a spatial (or in one or two instances temporal) sense as their basic meaning and that the other meanings are derived by metaphorical extension. Although the theory is intended to extend eventually to all prepositions, spatial, and non-spatial, here I will concentrate on the locative prepositions and, in particular, those which deal with the space around the reference object or place. On the basis of the place cell model, I will categorize the spatial prepositions into those that locate the vector components and those that locate aspects of the gaussian components or the boundaries of the field. When speaking of the strictly spatial meaning of the prepositions, I will refer to the area under the intersection of the gaussians as the **place field**. In the context of the broader non-spatial meanings of the prepositions, I will refer to this area as the **semantic field**.

2. A Computational Model for Place Representations

Single neurons in a cortical structure in the rat brain called the hippocampus become active when the animal visits a patch of a familiar environment. Different **place cells** have different preferred patches or **place fields** in the same environment. Experiments in which the place fields of the same neurons were recorded in different-shaped rectilinear boxes suggest that each field is the composite of two or more subcomponents, each of which is fixed to a large feature of the environment such as a wall of the room or the holding box (O'Keefe and Burgess, 1996). On the basis of these experiments, we have suggested that each subcomponent of a place field might reflect a separate input coming from cortical structures afferent to the hippocampus. In subsequent work, we have generalized this model so that it applies to environments and features of different shapes and can be used to predict the response of each cell in a wide variety of environments (Hartley et al., 2000). These putative inputs are termed **Boundary Vector Cells (BVCs)**. BVCs fire as a function of the distance of the animal in a specific direction from a large environmental feature such as the wall of the holding box. The model states that each place cell has inputs from two or more BVCs. It is with different aspects of a generalized version of these putative BVC inputs that I wish to identify different spatial prepositions. The basic idea is that each of the spatial prepositions sets up a different spatial field and these are subsequently combined in the hippocampus to form complex structures that underpin the interpretation of episodes, narratives, and narrative memories. The equation for the field of a putative BVC input is

(1) $\quad g_i(r,\theta) \propto \exp[-(r-d_i)^2/2\sigma_{rad}^2(d_i)]/\sqrt{2\pi\sigma_{rad}^2(d_i)}$
$\quad \times \exp[-(\theta-\varphi_i)^2/2\sigma_{ang}^2]/\sqrt{2\pi\sigma_{ang}^2}$

The receptive field is a function of four parameters:

(2) $\quad d_i, \varphi_i, \sigma_{rad}^2, \sigma_{ang}^2$

ϕ_i is the angle of the direction vector in polar coordinates relative to a reference direction. This reference direction can be determined in several different ways but in general is fixed to an environmental frame rather than to an egocentric frame;
d_i is the distance of the center of the field from the origin along the direction vector φ_i;
σ_{rad} is the width of the field in the radial dimension; while
σ_{ang} is the width in the angular dimension.

In the cognitive mapping system of the rat, the **location of the BVC field** is given by the length of the direction vector (i.e. a distance in a particular direction) from the animal itself (see Fig. 4.1).

The **size and shape of the BVC field** are given by the distance from that landmark in the radial direction and by a fixed angular width in the other. This results in the fields close to the rat being smaller and more strongly peaked than those farther away. In the human linguistic system, an analysis of the prepositions suggests that the fields are sometimes located relative to the

BVC Model

FIG. 4.1 Three examples of the fields of boundary vector cells. Each cell identifies a teardrop-shaped region whose center is located at a fixed distance and direction from the animal and whose shape depends on the distance from the animal. Fields farther away from the animal are broader and less sharply peaked (after Hartley et al., 2000).

speaker or listener but more generally relative to other objects in the environment (the reference landmarks). Furthermore it would appear that the size and shape of the semantic field can be specified independently of its distance from the reference landmark. As we shall see in the subsequent sections, some prepositions (e.g. *beyond* and *below*) identify only one of the radial borders of the field while others (e.g. *behind* and *under*) additionally specify the two angular boundaries, and still others (such as *between*, *among*, and *in*) delimit all four boundaries. These latter can the thought of as formed by the intersection of several simple Boundary Vector fields and as much closer to the fully-fledged place fields found in the hippocampus of the rat. In the cognitive mapping system of the rat, the direction vector is usually given by environmental information or internal signals derived primarily from the vestibular system to orient the sense of direction. In humans, the direction vector is given by the universal gravitational signal in the Z direction and is calculated in the horizontal XY plane by reference to two or more objects in the environment. Often one of these objects is the speaker or listener (i.e. a deictic origin). In the rat mapping system, the source of distance information is not known for certain but probably involves the integration over time of path integration signals derived from the animal's movement or the use of exteroceptive (e.g. visual) cues from familiar reference objects. In the human vector grammar system, distances are sometimes given by modifiers of the prepositions. These can be given as relative distances such as *farther than* or absolute distances such as *3 ft.*

(3) John was farther behind the house than Jim.
(4) John was 3 ft. behind the car.

In general these modifiers act upon the length of the vector specified by the preposition, either in comparison to another vector or by specifying an absolute length. It does not seem possible to modify the other parameters specified by the prepositions such as the width or length of the place field.

3. The Semantics of Selected Prepositions

As examples of this approach, I will discuss the prepositions *beyond, below, behind, under, by, beside*, and, finally *between, in*, and *among*.

3.1. *Beyond* and *Below*

In O'Keefe (1996) I suggested that the meaning of *beyond* could be captured by the set of vectors whose length bore a specific relationship to a reference vector drawn to the distal surface of the reference object. Specifically it consisted of all vectors whose inner product when projected onto the direction reference has a larger magnitude than a reference vector drawn from the origin to the far end of the reference location or object. No restrictions were placed on the radial or angular width of the terminations of the set of vectors specified. Usually these would be expected to be larger than the width and length of the reference object. While the proximal boundary of the vector field was well specified, the distal boundary was left open. **Modifiers** such as *just* or *far* place restrictions on the length of the semantic field vector in relation to the distance to the landmark.

(5) The barn is *just beyond* the lake.
(6) The barn is *far beyond* the lake.

The role of the comparative modifier *farther* is to increase the distance δ and thus the location of the semantic field. It also has the subsidiary effect of placing an upper limit on the distal boundary of the secondary semantic field (in the example below, that of the house).

(7) The barn is *farther beyond* the lake than the house.
$\delta_b > \delta_h$

Below has a meaning analogous to *beyond* in the vertical direction. In O'Keefe (1996) I gave its meaning as the field of vectors whose distance along the vertical reference vector was more negative than that of the reference object.

One of the problems with this use of a vector field to designate the place fields located by prepositions is that it describes a homogeneous field within which the locandum has an equal probability of being found at every location. While this may be an accurate depiction of the semantic field of *beyond* for

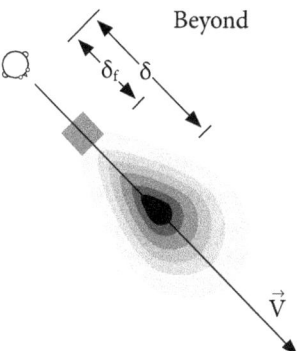

FIG. 4.2 The place identified by the preposition *beyond*. The area is shown as a set of grayscale concentric contours, the inner, darker colors indicate a higher probability of finding the object to be located. The center of the field is located by a point along the direction vector V at a distance δ from the origin (in this example the Observer). Distance δ is greater than $δ_f$ the distance from the origin to the farthest edge of the reference object (shown as a grey block). The length and width of the place field are left unconstrained and the width may be larger than the width of the reference object.

some speakers of English, it does not accord with my own intuitions, which suggest some internal structure to the fields of *beyond* and *below* such as that represented in Figs. 4.2 and 4.3. This intuition is even stronger in the case of other related but more complex prepositions such as *behind* and *under* (see below).

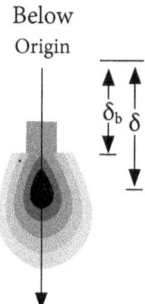

FIG. 4.3 The place identified by the preposition *below* has its center on the vertical direction vector drawn through the reference object at a distance $δ > δ_b$, the lower edge of the reference object. Like the field of *beyond*, there is no restriction on the width of the field in the orthogonal horizontal plane.

It is obvious that the structure of the semantic fields shown in Figs. 4.2 and 4.3 bears a strong resemblance to those of the BVC cells in Fig. 4.1. One interpretation of the semantic field idea is that it describes the probability of finding the located object within the field, in this case highly likely close to the center and less likely as one moves towards the periphery. Empirical support for the idea that there is structure to the space identified by prepositions such as *below* comes from the work of Logan and Sadler (1996). They showed students two letters on a video screen, a centrally placed O and an X at one of 49 locations, and asked them to rate how acceptable on a scale of 1 (least acceptable) to 9 (most acceptable) the location of X was as an exemplar of the sentence 'the X is [relation] the O'. Among the relationships tested were *above* and *below*. The average acceptance profiles for each location of X for the relations *below* (A) and *above* (B) are plotted as contour maps in Fig. 4.4.

While Logan and Sadler divide the fields into three regions, good, bad, and acceptable, in line with their **spatial template model**, it is clear that the field might be better described as a continuous region peaked in the center and falling off gradually in a monotonic fashion in all directions. These fields look remarkably similar to the teardrop-shaped fields shown in Fig. 4.3. Further studies of this nature will be necessary to identify the exact mathematical

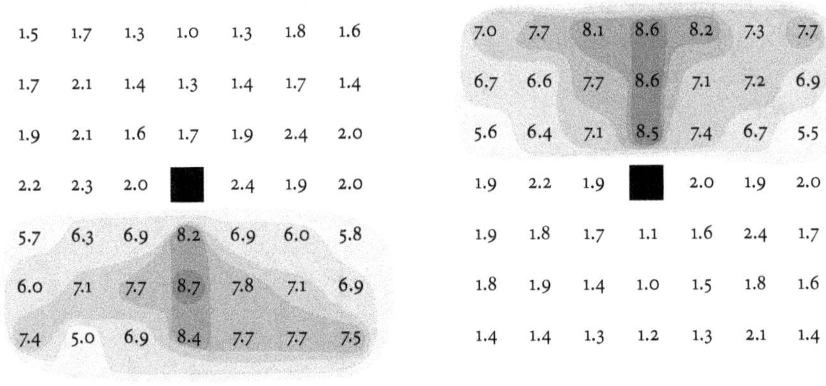

1.5	1.7	1.3	1.0	1.3	1.8	1.6		7.0	7.7	8.1	8.6	8.2	7.3	7.7
1.7	2.1	1.4	1.3	1.4	1.7	1.4		6.7	6.6	7.7	8.6	7.1	7.2	6.9
1.9	2.1	1.6	1.7	1.9	2.4	2.0		5.6	6.4	7.1	8.5	7.4	6.7	5.5
2.2	2.3	2.0	■	2.4	1.9	2.0		1.9	2.2	1.9	■	2.0	1.9	2.0
5.7	6.3	6.9	8.2	6.9	6.0	5.8		1.9	1.8	1.7	1.1	1.6	2.4	1.7
6.0	7.1	7.7	8.7	7.8	7.1	6.9		1.8	1.9	1.4	1.0	1.5	1.8	1.6
7.4	5.0	6.9	8.4	7.7	7.7	7.5		1.4	1.4	1.3	1.2	1.3	2.1	1.4
(a)								(b)						

FIG. 4.4 (a): Acceptability ratings for locations in terms of how well they satisfy the relationship *below* the O (marked by the grey square here). Higher numbers denote greater acceptability. Contour maps have been superimposed on the lower half of the figure to provide an idea of the semantic field of the preposition *below*. (b): Acceptability ratings for the preposition *above*. Contour levels used in both plots are 8.5, 7.5, 7.0. 6.0, and 5.0. Modified from Logan and Sadler (1996) with permission. See original paper for additional details.

functions associated with each preposition but the general correspondence to the proposed model is encouraging.

3.2. *Behind* and *Under*

In O'Keefe (1996), I defined *behind* as the set of vectors with a larger magnitude then a reference vector but with an angle less than or equal to the vector drawn to the outer edge of the reference object. Within the current framework, *behind* is defined as shown in Fig. 4.5.

The field is the same as that for *beyond* except that the width is restricted by the angle φ. This is the angle made by the vector drawn from the origin to (i.e. along) the outer edge of the reference object. The field structure is such that there is a greater likelihood of finding the locandum towards the center of the field and less likelihood as one moves towards the edges.

The preposition *under* was previously viewed as having two meanings; one similar to that of *below* and a second slightly more restricted one. This second meaning differed from that of *below* in that it further restricted the region in the horizontal plane to the projection of the reference object onto that plane. In the model presented in this chapter, the region of *below* (and $under_1$) is

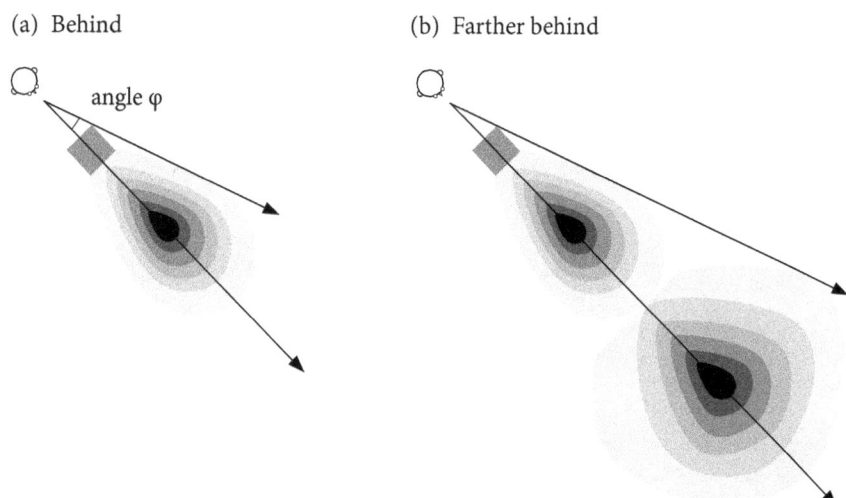

FIG. 4.5 (a): The place identified by the preposition *behind* is the same as that of *beyond* except for greater restrictions on the axial width of the field. The width of the field is determined by the vector with angle φ to the direction vector. (b): The semantic field of *farther behind*.

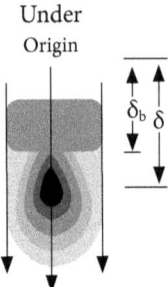

FIG. 4.6 The place identified by the preposition *under* is the same as *below* except that the width of the field in the horizontal plane is restricted by vectors drawn parallel to the vertical reference vector through the outermost edges of the reference object.

identified as shown in Fig. 4.3 and that of *under₂* in Fig. 4.6. The center of the field of *under₂* is given by the distance δ_d from the origin of the **vertical direction vector** and δ_b, the distance from the edge of the field. The boundaries of the field and the extent of the region *under* in the XY plane are determined by vectors parallel to the vertical direction vector and intersecting the sides of the reference object.

The importance of distances in the horizontal XY plane is illustrated by the operation of the modifier *farther* in this dimension (cf. Fig. 4.7b).

(8) The white box is farther under the shelf than the white circle.

Notice that the interpretation of this sentence depends on whether the shelf is located against a wall or not. When the reference object is such that it can

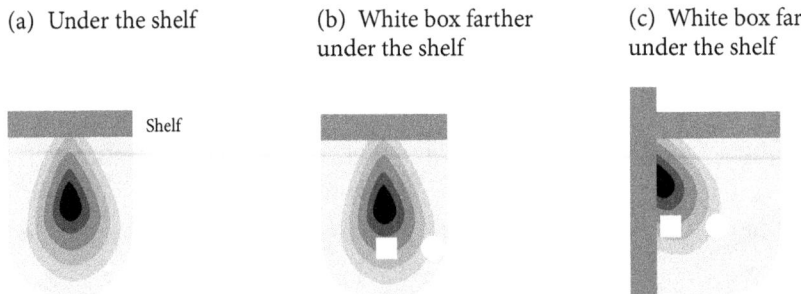

FIG. 4.7 (a) shows the field of *under* the horizontal shelf. In both (b) and (c), the box is farther under the shelf than the sphere.

be approached from either side then the location of the peak of *under* would appear to be determined by the center of the shelf. If, however, approach is only possible from one end, as in Fig. 4.7c where the shelf is fixed to a wall, then the peak of the field is close to the wall. This is a good demonstration that the field need not be symmetrical nor need the field peak be located in the geometrical center of the field.

The action of comparators such as *farther* in Fig. 4.7 is different from the effect of these modifiers on the preposition *below* (and $under_1$ when it is used with reference objects which have very large or unlimited extents in the horizontal plane) where they appear to act on the length of the vector δ in the vertical direction.

(9) Farther below the surface of the lake (under the water).

It is of interest that the subjects in the experiments of Logan and Sadler (American college students) did not distinguish between *below* and *under*, suggesting that they were restricting their usage to the first meaning of *under*. This may have been due to the use of relatively small letters as the objects to be located in this experiment. It would be interesting to see if a distinction between *below* and *under* emerges when horizontally extended reference shapes are used.

3.3. *By* and *Beside*

By is shown in Fig. 4.8. The region designated by *by* or part of this region can also be identified by the preposition *beside*. Whether the whole of the *by* field, or only part of it, is considered to be *beside* the reference object depends on

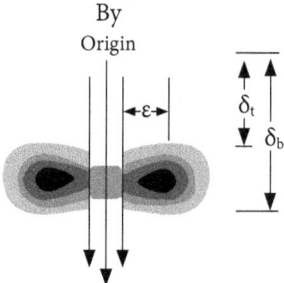

FIG. 4.8 The place identified by the preposition *by* is a toroid-like region whose center is located between $δ_t$ and $δ_b$ at a distance ε from the outer edges of the reference object.

the availability of a reference vector in the orthogonal XY plane to polarize the region into *before* and *behind* subcomponents. In the absence of a horizontal reference vector, the entire region is *beside* the reference object. When it is present, the region designated by *by* can be further subdivided into four continuous regions, one in *front of* or *before* it, one *behind* it (see above), and two *beside* it. The *beside* field is restricted to that part of the *by* region which is neither *before* nor *behind* the reference object. This means that *beside* involves a third-order computation, depending on the prior identification of the *above* and *below* regions followed by the *front* and *back* regions. This may explain why the speed with which these judgements can be made about the location of an object are fastest for *above* and *below*, followed by *front* and *back*, and slowest for *left* and *right* (beside) (see Tversky, 1996). This analysis would seem to suggest that an object cannot be both behind and beside a reference object at the same time.

(10) *The car beside and behind the house

For a small number of objects (most notably humans, animals, furniture), the usual orientation of the reference vector leads to their being assigned a front, back, and sides, especially when they are in their canonical orientations. This has led to the idea that these locations are determined by the **major axes of the object** (Landau and Jackendoff, 1993). This might tempt us to think that these prepositions refer to space that is fixed to the objects themselves. However, an alternative interpretation is that, in appropriate contexts, the axes of the object determine the orientation of the direction vectors. For example, for a human this would mean that the **canonical direction vector** for *behind* would have an orientation pointing from the person's chest to his or her back. However, these canonical interpretations are easily overridden in non-standard situations, for example, when the reference object is in an unusual orientation.

(11) the empty bottle in front of the upturned chair
(12) Stand beside that man lying on the floor.

It should be further noted that the ability to identify canonical orientations depends on the availability of vertical, and in some cases horizontal, reference vectors in the first place.

3.4. *Between*, *In*, and *Among*

These are the prepositions that come closest to defining fully-fledged place fields. They specify not only the location of the field center but also its boundaries on all sides. They differ in the way in which the field boundaries are speci-

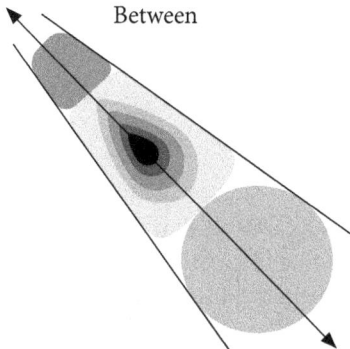

Fig. 4.9 The semantic field of *between* the square and the circle.

fied. *Between* specifies two reference objects which determine the boundaries, *among* three or more objects which locate the perimeter of the field, while *in* identifies an object or a location the boundary of which coincides with that of the field. Figure 4.9 shows the field of *between*. The center of the field is located on the line joining the centers of the two reference objects and the boundaries are formed by (1) the surfaces of these objects facing each other and (2) by the lines connecting the outer surfaces of the reference objects. The field structure suggests that objects located closer to the line joining the midlines of the two reference objects will be perceived as better instantiations of *between* than those closer to the periphery.

The field of *in* is bounded by the inner surface of the reference object with a peak somewhere near the centroid. Note that the field of *in* need not necessarily be coextensive with the reference object that defines its boundary, in particular where the reference object is not a continuous one. For example, the field of a bowl that is open at the top may continue to follow the extrapolated curvature of the bowl and not the object itself and objects located in this region would still be *in* the bowl.

3.5. Synecdoche and the Semantics of the English Prepositions

The strong geometric interpretation of the prepositions taken in this chapter would seem to invite the criticism that it gives an overly precise interpretation of their meanings. It should be remembered however that the place fields represent probability distributions within which objects may be found and do not specify precise locations as such. Furthermore, my view is that some of the exceptions and ambiguities that might be used to challenge this view can be answered by a closer look at the way in which synecdoche and ellipsis influence the interpretation of prepositional phrases.

(13) flowers in the vase
(14) chair under the table

Both phrases clearly represent the use of the whole for the part. In the first sentence it is the stems of the flowers that are in the vase, while in the second sentence it is usually the seat of the chair that is under the horizontal surface of the table (and not e.g. the back). I suspect that some of the functional interpretations of the meanings of the prepositions may yield to a similar analysis.

Summary

1. **Cognitive Map Theory** suggests that the function of prepositions is to locate place fields within a Semantic or Narrative Map, the linguistic analog of a spatial map.

2. **Semantic Maps** are devices for storing the relationships between linguistic entities such as nouns in the form of narratives, including the personal narratives or stories that conscious subjects construct about their experiences.

3. In previous work I have suggested that entities within the map were located by sets of vectors. Here I offer a more natural representation in terms of **place fields** analogous to those found in the hippocampus of the rat. These fields are identified by their center and spread in two or three orthogonal dimensions. **Simple semantic fields** resemble those we have modeled as Boundary Vector Cells. They are located at the end of a vector and have relatively fixed boundaries that depend on the distance from the reference object. More **complex semantic fields** are closer to true place fields and may be located by the intersection of two or more simple fields, each determined by a vector in a different direction. Unlike the previous vector field theory, which gave a uniform probability of locating an object in an area, the current place field theory provides internal structure to the area located by the preposition and suggests that the target to be located is more likely to be found in certain parts of the field than in others.

4. **Field Centers** are located at the termination(s) of one or more principal vectors. The orientation of the principal vertical direction vector is usually provided by the universal gravity signal. On the other hand, horizontally oriented vectors need to be computed on the basis of a vector orthogonal to the horizontal vector or to the relative positions of landmarks, usually by vector subtraction of the vectors from the observer to two of these landmarks.

5. Some prepositions (e.g. *beyond*, *below*, and *above*) identify regions that are relatively unrestricted in the plane orthogonal to the primary location vector. For this class of prepositions, comparative modifiers such as *farther* or *near* operate to increase or decrease the length of the principal vector. Other prepositions such as *behind*, *by*, and *beside* place further restrictions on the boundary of the field. Prepositions such as *between*, *in*, and *among* identify the boundaries of the field completely.

5

The Unique Vector Constraint: The Impact of Direction Changes on the Linguistic Segmentation of Motion Events

JÜRGEN BOHNEMEYER

Abstract

This chapter explores a principle of motion event coding that appears to be shared across languages: the unique vector constraint (UVC). The UVC determines the complexity of direction information that can be coded in simple motion event clauses. According to the UVC, all direction specifications in a single simple clause must denote the same direction vector. The UVC is distinguished from another principle of motion event coding, the argument uniqueness constraint (AUC), which requires the unique assignment of path roles such as 'source', 'via', 'goal', 'toward', etc. in single simple clauses. It is argued that both constraints are principles of form-to-meaning mapping at the syntax–semantics interface.

During the preparation of the paper that forms the basis of this chapter, I benefited tremendously from comments, suggestions, criticism, and examples provided by F. Ameka, B. Bickel, M. Bowerman, P. Brown, A. Cutler, N. Duffield, S. Eisenbeiß, N. Enfield, J. Essegbey, S. Kita, W. J. M. Levelt, S. Meira, B. Narasimhan, E. Schultze-Berndt, A. Senghas, D. I. Slobin, M. Swift, A. Terrill, D. P. Wilkins, and the editors of this volume E. van der Zee and J. Slack. The evolution of the central idea, the unique vector constraint, proceeded to a significant extent as an extended conversation between me and several of these colleagues, and the proposal would have never reached the stage at which it is presented here without their feedback. I hasten to acknowledge that some of them will still disagree with the idea. Of course, all errors and shortcomings of this chapter are exclusively my responsibility. I would also like to thank the attendants of the Lincoln workshop and of several presentations at which I had the opportunity to discuss parts of the content of this chapter, at Amsterdam University, MPI Nijmegen, the University of Hawai'i at Mānoa, and the University of Rochester.

Introduction

This chapter draws on ongoing research of the Event Representation project at the Max Planck Institute for Psycholinguistics in Nijmegen. The Event Representation group is dedicated to investigating universals and cross-linguistic variation in the linguistic representation of complex events. Motion is one of three domains in which members of the group have been studying the coding of complex events. The results of a pilot study conducted in 1999 on thirteen mostly unrelated languages from six continents show a striking amount of variation in the coding of locomotion events. Section 1 introduces the stimuli that have been used in this investigation and illustrates variation in the representation of **motion paths** (in the sense of Jackendoff 1983 and Talmy 1972, 1985, 1991) with examples from Yucatec Maya. In this language, path is exclusively lexicalized in a set of verbs of 'inherently directed motion' (Levin 1993); there is no differentiation whatsoever of source, goal, or location outside these verb roots. Therefore, a movement from source to goal has to be distributed across a minimum of two mutually independent clauses in Yucatec, one referring to a departure event and one referring to an arrival event.

However, the group also found apparent non-trivial universals in the coding of complex motion events. Of central concern in this chapter is a constraint on the information about the direction of the moving object that can be accommodated in single motion event clauses. In sect. 1, data are presented that motivate the assumption that such a constraint is operative in English and other languages. The crucial question to be addressed here with respect to these data is how the constraint assumed to underlie them is to be formulated. In sect. 2, a parallel is drawn between this constraint and another one that has also emerged from the Event Representation data. The **'argument uniqueness constraint'** (AUC) requires path argument roles to be uniquely mapped onto Ground-denoting expressions within single clauses. This is an apparently universal constraint on the coding of **motion events** at the syntax–semantics interface. It is argued in sect. 3 that the AUC and the constraint on motion direction are independent principles of the same kind, namely, principles of form-to-meaning mapping. Thus, it is shown that the data on motion event descriptions under direction changes can be accounted for only in terms of the direction information that can be *coded* in single clauses, not (directly) in terms of the *extensional* trajectory traversed by the moving object. In order to capture relations across directions, the notion of direction vector is introduced. Based on this notion, a statement of the **unique vector constraint** (UVC) is proposed.

Section 4 addresses the role of world knowledge in the coding of motion events. It is shown how linguistic representations that do not entail changes in direction may implicate such changes in particular contexts, and hence can be used in these contexts to refer to motion events in which direction is not preserved. Section 5 takes a brief look at the linguistic coding of path shape. It is argued that path shape and direction vectors are, in first approximation, independent components of motion paths, and that the coding of both these components is constrained by similar but independent principles.

1. The Impact of Direction Changes on the Coding of Motion Events

In a 1999 pilot study by the Event Representation group, the so-called ECOM clips (short for Event Complexity) were employed, a set of computer animations that showed scenarios varying according to parameters expected to trigger cross-linguistic variation in macro-event construal. In the context of this chapter, two subsets of the ECOM clips are of particular interest. These are first clips in which complexity is increased from scene B1 to scene B5 by gradually adding further **Ground objects**, while the trajectory traversed by the moving **Figure** (Talmy 1985; Landau and Jackendoff 1993) and the orientation of the latter are kept constant (ECOM B1–B5); and secondly clips in which the same configuration of potential Ground objects is constant across scenes, but the Figure's trajectory across these Grounds is gradually extended by adding new segments with every consecutive clip (ECOM C1–C6). Crucially, with each additional segment of the trajectory, a 45- or 90-degrees change in the direction of the Figure is introduced. The objects involved in these scenes are simple geometrical shapes (circles, squares, triangles, etc.). In the most complex scene of set B, B5, there are four Ground objects, a square at the beginning of the motion event, a triangle at the end point, and a bar and a house-shaped object in between. Figure 5.1 shows the first frame and the final frame of B5. Figure 5.2 shows the first and the final frame of C6, in which the circle (the Figure) rolls up the inside of a U-shaped container, exits the container, descends on its outside, rolls over to the triangle, and ascends to the triangle's top.[1]

FIG. 5.1 First and last frame of ECOM B5.

FIG. 5.2 First and last frame of ECOM C6.

[1] It may be noticed that there are small indentations in the circle and square. These served as negative 'extremities' for grabbing and holding objects in non-motion scenes.

In approaches to the linguistic representation of motion events such as Jackendoff (1983: ch. 9) and Talmy (1972, 1985, 1991), it is implicitly assumed that scenarios such as those shown in ECOM B1–B5 can be represented in single simple clauses in all languages, just as they can in English. This is not the case. The languages sampled in the ECOM study form a cline in terms of the scenarios they can express in single simple clauses. Some languages can indeed, like English, code all five scenarios in this way. For example, in Dutch (investigated by the author and M. Caelen), it is entirely possible to describe B5 in a single clause, as in (1):

(1) Het balletje rolt van het vierkant over een baan voor het huisje langs
    the little ball rolls from the square along a track past the little house
    naar het driehoekje.
    to the little triangle
    'The little ball rolls from the square along a track past the little house
    to the little triangle.' (ECOM D B5 constructed)

Notice, however, that Dutch subjects *preferred* not to mention more than two Ground objects per clause in their descriptions, as in (2):

(2) Een rood rondje komt van de linkerkant waar een blauw dingetje
    a red round thing comes from the left side where a blue thing
    staat, rolt dan naar de rechterkant over een baan heen,
            rolls then to the right side   across a track
    maar het lijkt nu een beetje straat, want   er staat   een huisje
    but it looks now a bit like a street, because there stands a house
    achter. Komt tot stilstand   tegen een groen driehoekje.
    behind comes to a stand still against a green triangle
    'A red round thing starts from the blue thing on the left, then
    rolls to the right across a track; but that actually looks a bit like
    a street, because behind it there's a house. It stops at a green
    triangle.' (ECOM D B5 S3)

At the other end of the cline are languages such as Yucatec Maya (studied by the author) which cannot express more than one location-change event in a single clause. As argued in more detail in Bohnemeyer (1997; submitted) and Bohnemeyer and Stolz (submitted), there is no lexicalization of path notions in Yucatec outside a small set of verb roots of 'inherently directed motion' (Levin 1993: 263) translating 'go', 'come', 'enter', 'exit', 'ascend', 'descend', etc. There is in particular no distinction of locative relations and **source** or **goal** relations in Ground-denoting expressions. Consequently, a locomotion leading from source to goal has to be broken down into a minimum of two clauses, one representing the departure and one representing the arrival. Thus, (3) is a natural rendition of 'Pedro went from X-place to Y-place' in Yucatec:[2]

(3) Pedro-e', ti' yàan t-u kàah-il X, káa h bin-ih,
Pedro-TOP LOC EXIST(B.3.SG) LOC-A.3 live-REL X CON PRV go-B.3.SG
káa h k'uch t-u kàah-il Y.
CON PRV arrive(B.3.SG) LOC-A.3 live-REL Y
'Pedro, he was in X-place, (and/then) he left, (and/then) he arrived in Y-place.' (constructed)

Consequently, a description of ECOM B5, which involves a source, a goal, and two **via** (i.e. 'mid-way') Ground objects, has to be distributed across a minimum of four clauses in Yucatec. Example (4) is a Yucatec description of B5:[3]

(4) Ba'l–e', be'òora-a' t-inw il-ah-e', hun-p'éel chan
thing-top now-D1 PRV-A.1 see-CMP(B.3.SG)-TOP one-CL.IN DIM
áasul ba'l k-u p'áat-al t-u xùul le tu'x h
blue thing IMPF-A.3 await\ACAUS-INC LOC-A.3 end DEF where PRV
luk' le chan ba'l chak-o', k-u bin u balak'-e',
leave(B.3.SG) DEF DIM thing red(B.3.SG)-D2 IMPF-A.3 go A.3 roll-TOP
k-u tso'k-ol-e', k-u máan y-iknal hun-p'éel chan ba'l
IMPF-A.3 end-INC-TOP IMPF-A.3 pass A.3-at one-CL.IN DIM thing
chak xane', k-u tso'k-ol-e', k-u k'uch-ul
red(B.3.SG) also IMPF-A.3 end-INC-TOP IMPF-A.3 arrive-INC
y-iknal le triàangulo áasul-o'.
A.3-at DEF triangle blue(B.3.SG)-D2
'But, this time, I saw a blue thing, it remains at the end where the red thing left, [the red thing] went rolling, then it passes by a thing which is also red, then it arrives at the blue [i.e. green] triangle.' (ECOM Y B5 RMC)

---

[2] Abbreviations in interlinear morpheme glosses include the following: 1 = 1st person; 3 = 3rd person; A = cross-reference set-A ('ergative'/'possessor') clitics; ACAUS = anticausative derivation; B = cross-reference set-B ('absolutive') suffixes; CL = classifier (numeral/possessive); CMP = completive 'status' inflection; COMP = comparative particle; CON = connective particle; D1 = proximal deictic particle; D2 = distal deictic particle; D4 = negation-final/locative particle; DEF = definiteness determiner; DIM = diminutive particle; EXIST = existential/possessive/locative predicate; HORT = hortative particle; IMPF = imperfective aspect marker; IN = inanimate (classifier); INC = incomplete 'status inflection'; LOC = generic preposition; POS = positional (verb stem class); PRV = perfective; REL = relational derivation (nouns); SG = singular; TOP = topic marker.
[3] Notice that this description in fact omits one of the Ground objects of ECOM B5, namely the yellow bar in between the square and the triangle. Despite the efforts to get the consultants to mention the entire spatial layout of the clips, such omissions occurred quite frequently.

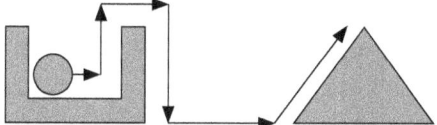

Fig. 5.3 Motion vectors in ECOM C6.

Intermediate positions on the cline of path integration are occupied by languages in which most (Japanese, studied by S. Kita) or all (Ewe, studied by F. Ameka and J. Essegbey) of the scenarios may be expressed by mono-clausal multi-verb constructions.

Of key interest for the purposes of this chapter is the striking difference in the distribution of descriptions elicited by the B clips and by the C clips. It was found that even those languages that allowed for monoclausal descriptions of the B clips forced a breakdown of descriptions of the C clips into multiple clauses. Hence, descriptions of the motion scenes in the ECOM C clips aligned much more across the various languages studied by members of the Event Representation group than descriptions of the ECOM B scenes. None of the languages permits the packaging of more than one of the trajectory segments of the ECOM C scenes per clause-level unit. Figure 5.3 gives a schematic representation of the segments in ECOM C6 as vectors (cf. Fig. 5.2 above).

Example (5) below is a description of ECOM C6 in Yucatec. Just as in the description of ECOM B5 reproduced in (4) above, one trajectory segment is being referred to (explicitly or implicitly) per clause, which in the case of C6 essentially amounts to a ratio of one clause per motion vector (however, two of the vectors of C6, or two of the 'legs' of the locomotion event—the ascension on the inside of the container object and the descending event on its outside—are not mentioned at all):

(5) Ich le   chan kwàadrado yàan   hun-p'éel chan sìirkulo
    in DEF DIM square   EXIST(B.3.SG) one-CL.IN DIM circle
    chak-i'.   Kóoh-ol   u   tàal u   balak'-e',
    red(B.3.SG)-D4 hit\ACAUS-INC A.3 come A.3 roll-TOP
    k-u   hóok'-ol ich le   kwàadrado áasul-o',
    IMPF-A.3 exit-INC in DEF square   blue(B.3.SG)-D2
    k-u   séegir u   balak'-e',
    IMPF-A.3 continue A.3 roll-TOP
    k-u   k'uch-ul   tak te   hun-p'éel chan triàangulo-o',
    IMPF-A.3 arrive-INC even LOC:DEF one-CL.IN DIM triangle-D2
    ko'x a'l-ik   hun-p'éel chan piràamide.
    HORT say-INC(B.3.SG) one-CL.IN DIM pyramid

K-u   ẏna'k-al   tak   t-u   máas ka'nal le   chan piràamide,
IMPF-A.3 ascend-INC even LOC-A.3 COMP high DEF DIM pyramid
ti'   k-u   na'k-al   pek-tal-i'.
there IMPF-A.3 ascend-INC sit.on.surface-POS.INC-D4
'In the square there is a red circle. It comes rolling hitting [making contact with the square], it exits the blue square, it keeps rolling, it arrives at a triangle, let's say a pyramid. It ascends to the highest [point] of the pyramid, there it ascends to rest.'   (ECOM Y C6 EMB)

Now compare this to a Dutch description of the same scene:

(6)   Aan de linkerkant van het scherm zit een blauw kokertje of een
      on the left side   of the screen   sits a blue case   or
      bakje. Er zit   een rood balletje in,   en   dat rode balletje gaat
      box. There sits a red ball   inside and that red ball   goes
      rechtsom via de rechterkant van het kokertje of het bakje naar
      rightwards via the right side of the case   or box   to
      buiten,   over de top   naar beneden, en   rolt dan naar rechts
      the outside over the top to the ground and rolls then to the right
      richting een groen driehoekje wat   daar ligt, en het balletje rolt
      towards a green triangle   which there lies and the ball   rolls
      tegen   het driehoekje op naar boven   en komt op de
      against the triangle   up to the upper side and comes on the
      bovenkant van het groene driehoekje tot stilstand.
      upper side of the green triangle   to a standstill
      'On the left of the screen there's a blue case or box. There's a red ball inside, and that ball goes to the right along the right side out of the case or box, over the top to the ground, and rolls to the right towards a green triangle, and the ball rolls up along the side of the triangle to the top, and it stops at the top of the triangle.'   (ECOM D C6 S3)

In (6), changes in the direction of the Figure are either left unmentioned (e.g. 'the ball goes rightwards along the right side out of the box', neglecting the change from horizontal to upward motion), or they lead to the insertion of a clause boundary. The structurally minimal solution to break down the description in case a change in direction is mentioned is a gapping construction without overt coordination ('it goes out of the box, over the top to the ground'). Overt coordination with gapping ('(...) and rolls to the right') or without gapping ('(...) and the ball rolls up along the side of the triangle to the top') occurs as well.

The same changes in the Figure's direction in the stimulus lead to breakpoints in linguistic descriptions of the stimulus in Dutch, Yucatec, and all the other languages in the sample (if they are mentioned!). This holds irrespective of how many location changes with respect to consecutive Grounds can be integrated in single motion event clauses in the particular language under preservation of the Figure's direction. This is a surprising finding that demands explanation.[4]

2. A Role Model for the Statement of the Constraint on Direction Packaging: The Argument Uniqueness Constraint

Given the amount of variation shown above to obtain across languages in the segmentation and packaging of motion events, which is unpredicted from current theoretical and typological approaches to the coding of motion in language, it appears all the more significant that the research of the Event Representation group identified some intriguing candidates for non-trivial universals of motion coding. One such hypothetical universal, the 'argument uniqueness constraint' (AUC), is briefly discussed in the present section, because it illustrates the *type* of constraint that is assumed also to underlie the commonality found across the languages of the sample in the coding of motion events involving changes in direction.

The AUC has a scope much wider than the coding of motion events. It essentially states that no two structural arguments or adjuncts of the same clause can be assigned the same semantic role. In syntactic theory, this constraint is known under labels such as 'theta criterion' (in GB) or 'biuniqueness condition' (in LFG), but it is usually considered only with respect to syntactic core arguments (but see the remarks below on Nikanne 1990), which are realized as subjects and direct and indirect (or 'primary' and 'secondary') objects, and assigned 'case roles' such as 'agent', 'theme', 'recipient', etc. In this sense, the constraint was originally proposed by Fillmore (1968: 21):

The sentence in its basic structure consists of a verb and one or more noun phrases, each associated with the verb in a particular case relationship. The 'explanatory' use of this framework resides in the necessary claim that, although there can be compound instances of a single case (through noun phrase conjunction), each case relationship occurs only once in a simple sentence.

However, it seems clear that the constraint ruling out multiple assignments of the same argument role in the same simple clause has a very general scope. It

---

[4] The entire set of data collected during the ECOM pilot study is still being analyzed by members of the group. The results will be published in due course.

also holds for example for instruments, as noted already by Fillmore. In the same way, it is not possible to have more than one Ground-denoting phrase with the same path role in one and the same simple motion event clause. Consider, for example, a scenario in which the Figure starts out in the library, consecutively moves across the hall past the canteen and the reception to the entrance, and eventually leaves the building (this trajectory happens to match the spatial layout of the Max Planck Institute in Nijmegen):

(7) a. *Sally walked out of the library from the reception to the entrance.
 b. Sally left the library and walked from the reception to the entrance.
 c. Sally walked out of the library, from the reception to the entrance, and left the building.

Example (7a) is ungrammatical on account of assigning the source role twice among the semantic arguments of one verb, whereas (7b) and (7c) are fine. Example (7a) shows that the constraint is not of a purely structural nature; that is, it does not merely concern multiple uses of the same preposition (or syntactic relation). Notice also that the gapping construction in (7c) does not violate the AUC. Such elliptical constructions behave with respect to argument assignment like multi-clausal structures; they do not instantiate the 'simple' clauses the AUC is restricted to. The examples in (8) illustrate the same points made above for source specifications with respect to goal specifications:

(8) a. *Sally walked across the hall to the entrance out of the building.
 b. Sally walked across the hall to the entrance and left the building.
 c. Sally walked across the hall to the entrance and out of the building.
 d. Sally walked across the hall to the entrance, out of the building, and onto the parking lot.

The coordination in (8c) can, arguably, be analyzed in two ways, as a coordination of prepositional phrases yielding one internally complex goal phrase, or, more likely, as an underlying multi-clausal gapping construction (*...and walked out of the building*). Either way, (8c) does not violate the AUC.

Interestingly, English appears to distinguish a number of different via roles (contrary to what is apparently assumed in Jackendoff 1983), i.e. referential Grounds in the function of being passed by during a motion event, such as expressed by *along, across, through, over, past, by, via*, etc. Some via phrases can be combined in simple clauses, and thus apparently are assigned different path roles, while other via phrases are excluded from co-occurrence in single simple clauses:

(9) a. Sally walked across the hall past/via the canteen to the entrance.
   b. *Sally walked across the hall by the reception to the entrance.
   c. *Sally walked past the canteen by/via the reception to the entrance.

No violation of the AUC has been attested in any of the languages studied by the members of the Event Representation group. This therefore seems a plausible candidate for a universal constraint on motion event coding (and in fact, on event coding in general).[5]

In the remainder of this chapter, the AUC serves as a model for the introduction of the principle proposed to account for the segmentation of motion descriptions under changing direction. It is argued that these principles are two of a kind: they are neither purely formal nor purely semantic or conceptual restrictions, but restrictions on the linguistic coding possibilities at the syntax–semantics interface. An alternative generalization, equal in scope to the AUC, is proposed in Nikanne (1990: 30–1, 60–1). Nikanne suggests wellformedness rules on conceptual event representations in the framework of Jackendoff's (1983, 1990) Conceptual Semantics that exclude multiple applications of the same predicate function within such representations. Nikanne explicitly extends these rules to exclude multiple applications of the 'basic path functions' FROM, TO, TOWARD, AWAY-FROM and VIA. In contrast, in the present study, a level of semantic representation distinct from any non-linguistic mental representation is assumed, and the AUC and UVC are considered language-internal principles of form-to-meaning mapping—genuine *semiotic* constraints.[6]

---

[5] Goldberg (1991) rejects an account of the data presented here in terms of a general restriction on argument roles, and instead advocates a specific constraint on the expression of literal and metaphorical paths. She justifies this analysis with reference to cases such as *Sam tickled Chris off her chair silly*, where the result state expressed by the resultative construction might be considered a metaphorical goal.

[6] It is not prima facie obvious whether principles such as the AUC or the UVC obtain at the level of linguistic representations (as assumed here) or whether they are constraints on cognitive representations (as assumed in Nikanne 1990). The answer will depend to some extent on whether it is assumed that e.g. Dutch and Yucatec descriptions of ECOM B5 such as (1) and (4), respectively, encode identical conceptual representations at the level of 'conceptual structure' (CS). A positive answer to this question—suggested, perhaps, by the assumed universality of CS and the fact that something like (4) is the closest translation equivalent of (1) in Yucatec—would seem to discourage Nikanne's view (the constraint against multiple assignments of the same semantic role clearly does not hold at the level of multi-clausal discourse instantiated in (4)). Another relevant consideration may be modality. Initial evidence from Dutch Sign Language of the Deaf (*Nederlandse Gebarentaal*), collected by D. P. Wilkins with the ECOM clips, indicates that principles such as the AUC and the UVC are not valid for signed languages. If this is true, then Nikanne's proposal could only be maintained under the assumption that signed languages, unlike spoken languages, do not encode CS under 'representational modularity' (cf. Jackendoff 1990: 41–6).

3. Towards a Formulation of the Unique Vector Constraint

The argument uniqueness constraint discussed above is a restriction on what parts of a complex motion event can be expressed in one single simple clause. This constraint can be motivated in part from an analysis of descriptions of the ECOM B stimuli across various languages. Another restriction on the encoding of complex motion events, quite possibly equally universally valid, seems to affect the coding of motion scenes during which changes in the Figure's direction occur. Evidence for this constraint comes from the descriptions of the ECOM C clips presented in sect. 1. Example (10) seems a maximally explicit and maximally concise description of ECOM C6 in English:

(10) The ball rolls to the base of the inside wall of the container, then up the wall, over the top and out, down on the outside of the container, and on to the triangle and up to the top.

The sentence in (10) includes only one overt predicate. However, breaks in a typical intonation contour and the presence of coordinating conjunctions reveal (10) as comprising no less than seven clause-like units. Either the following assertions cannot be uttered under a single continuous intonation contour at all (i.e. they are not simple mono-clausal constructions), as is the case with (11c) and (11d), or they are not adequate descriptions of C6, as is the case with (11a) and (11b), *if* they are uttered as mono-clausal constructions:

(11) a. ?The ball rolls up (the wall of the container) over the top.
     b. ?The ball rolls up (the wall of the container) out (of the container).
     c. *The ball rolls up (the wall of the container) down (on the outside of the container).
     d. *The ball rolls down (at the outside of the container) up the triangle.

Examples (11a) and (11b) are possible descriptions of scenarios in which the top of the container is a slanted surface or in which the ball exits the container while going straight up, respectively. That is, they are acceptable as descriptions of scenarios in which the direction of the Figure does not change, which is not the case in ECOM C6. This shows that the constraint ruling out (11a) and (11b) as descriptions of ECOM C6 is, just like the AUC, neither a purely formal nor a purely semantic restriction, but one that limits the range of possible *interpretations* of simple clause structures, and therefore, a constraint on possible form-to-meaning mappings.

Again, with the introduction of coordination and/or gapping, the utterances in (11) become perfectly fine descriptions of ECOM C6:

(11) a'. The ball rolls up and (rolls/goes) over the top.
      b'. The ball rolls up and (rolls/goes) out.
      c'. The ball rolls up and (rolls/goes) down (on the outside of the container).
      d'. The ball rolls down and (rolls/goes) up the triangle.

These data suggests that there is a constraint on the clause-level packaging of descriptions of motion events involving changes in the Figure's direction. This constraint has been found valid across languages as typologically diverse in the way they code motion as Dutch (and English) and Yucatec. The question to be pursued in the remainder of this chapter is in what terms the constraint that rules out (11a–d), either absolutely or at least as valid descriptions of ECOM C6, should be stated. As a starting point, consider the scenario depicted in Fig. 5.4. A motion event depicted by the larger diagram in the upper left corner of Fig. 5.4 could be described in a single clause, as in (12):

(12) The Figure moved from A via B to C.

Example (12) describes the motion event in terms of **location change** only, specifying that A is the source and C the goal of the path, and that B is a via Ground. No information about the direction of the Figure at any point along the trajectory is revealed except for the entailments that the Figure moves

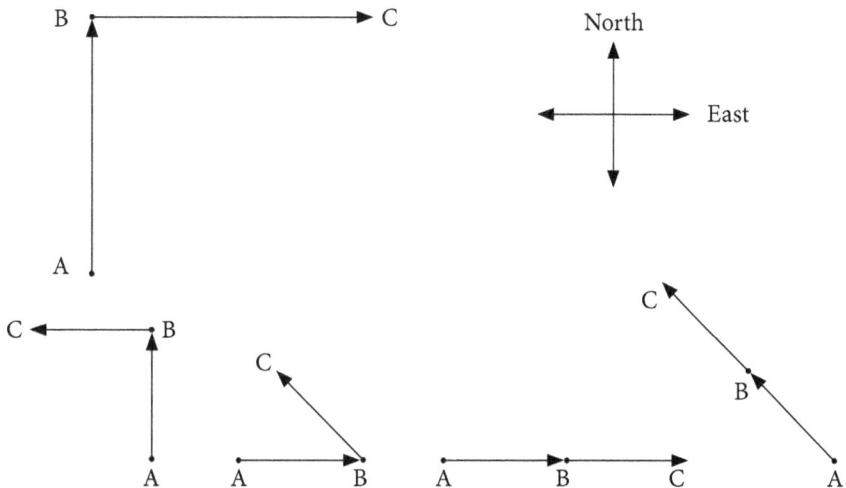

FIG. 5.4 The coding of direction in motion event descriptions.

away from A during appropriately small initial subintervals, towards C during appropriately small terminal subintervals, and towards and away from B, respectively, during appropriately small central subintervals. This does not entail any change of direction out of context: (12) may also serve as an adequate description of any of the motion events depicted in the smaller diagrams at the bottom of Fig. 5.4, including the one in which no direction change occurs. Now consider (13):

(13) a. *The Figure moved north via B east to C.
 b. *The Figure moved up via B left to C.
 c. The Figure moved north to B and then east to C.
 d. The Figure moved up to B and then left to C.

In (13), direction information (*north, east, up, left*) is coded in addition to location change in order to represent the motion event depicted in the upper left corner. This has the effect that the description becomes unambiguous in the context of Fig. 5.4. At the same time, it forces the use of coordination (or multiple independent clauses); cf. (13a) vs. (13c) and (13b) vs. (13d). The fact that the same scenario can be described in a single simple clause as long as only location change is coded but requires a more complex construction as soon as direction is specified indicates that the constraint at hand cannot be stated in terms of the extensional shape of the trajectory referred to in the motion event description. Different descriptions of the same extensional trajectory differ in their acceptability. It has been established in the discussion of (11) above that it is not the case either that the constraint operates exclusively on the adverbs and adpositional phrases that can be combined in single clauses. Therefore, the difference in acceptability between (12) and (13c–d) must depend on what information about the Figure's direction is asserted or entailed in the descriptions. The hypothesis to be advanced here, then, is that the constraint at issue affects precisely the information coded or entailed about the direction of a Figure in a single simple motion event clause.

The illformedness of (13a–b) can be accounted for by the AUC under the assumption that there are only two semantic roles that may be assigned to direction adjuncts or arguments. Such an assumption would follow naturally from Jackendoff's (1983) treatment of directions: Jackendoff holds that there are only two 'basic path functions' underlying direction specifications: TOWARD and AWAY-FROM. If these are translated into semantic roles, then (13a–b) would be illformed because they assign the toward role twice in a single simple clause. Nikanne's (1990) wellformedness rules on 'conceptual structures' also explicitly excludes event representations that apply the functions TOWARD and AWAY-FROM multiple times. But neither the AUC nor Nikanne's equivalent well-

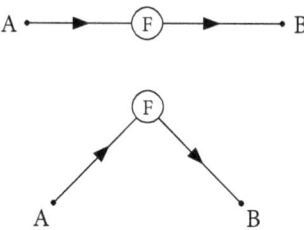

FIG. 5.5 The relationship between the AUC and the UVC.

formedness rules can account for the data presented in (11)–(11′). Consider Fig. 5.5. The AUC (or Nikanne's wellformedness conditions) cannot explain why (14a) is an adequate description of the motion event depicted in the upper diagram, but not of the event depicted in the lower diagram, while the opposite holds for (14b):

(14) a. The Figure moved away from A towards B.
b. The Figure moved away from A and then towards B.

Informally speaking, the unique vector constraint (UVC) proposed in this chapter to account for the data in (10)–(14) requires single simple clauses to specify no more than one direction, even if the direction information is encoded in multiple places in the clause, as in (14a). The remainder of this section attempts to achieve a more explicit formulation of this proposal, which entails clarification of what exactly is meant by direction and under what circumstances two direction specifications are considered to specify the same direction.

The characterization assumed here of how direction is encoded in language is adopted from Jackendoff (1983: 163–5). Jackendoff distinguishes three types of paths as represented in language: '**bounded paths**', '**routes**', and '**directions**'. All three are defined strictly relationally, i.e. with respect to referential Grounds. Bounded path Grounds define the beginning or end points of paths and are assigned source or goal roles, respectively. Route Grounds lie on the path in between source and goal; route functions are e.g. encoded by *via*, *past*, *through*, *across*, *over*, and *along*. Direction Grounds 'do not lie on the path, but would if the path were extended some unspecified distance' (Jackendoff 1983: 164). One diagnostic of direction specifications is that they do not entail location change. Therefore, motion clauses that contain only direction specifications, but not specifications of bounded paths or routes, are atelic. Consider the contrast in (15):

(15) a. Sally walked to her house in/*for five minutes.
b. Sally walked towards her house for/*in five minutes.

The bounded-path description in (15a) is telic and entails that Sally arrived at her house, while the direction description in (15b) is atelic and does not entail that Sally ever reached the house. The realization of the Ground varies with the Frame of Reference (FoR; cf. Levinson 1996); therefore, each FoR is associated with a set of direction expressions that are potentially unique to that FoR. Table 1 gives a few examples; (16) applies the telicity test to some of them.

(16) Sally went north/up/left for/*in five minutes.

TABLE 5.1 *Direction terms according to frame of reference*

Frame of reference	Absolute	Relative	Any (intrinsic/relative/absolute)	Combinations
Direction terms	*north(bound)*, *south(bound)* etc.; *up/down*; *upriver*, *downhill* etc.	*(to the) left(ward)*/ *(to the) right(ward)*	*towards (front/back/top/bottom etc. of) G/away from (...) G* ...	*up the hill*; *south towards the rock*; *to the left out of G*; ...

An interesting difference across the expressions listed in Table 5.1 is that only expressions relativized to Ground objects show two polarities: toward and away from the Ground object, respectively. This is a consequence of the polar nature of intrinsic FoRs. In absolute or observer-based FoRs, one cannot move 'away from the north/left', because *north* and *left* do not denote specific places (moving *away from the north (of England)* is of course fine, because *the north (of England)* does denote a place). Directions in absolute or relative FoRs are always defined with respect to the origin of the FoR; therefore, all the arrows in

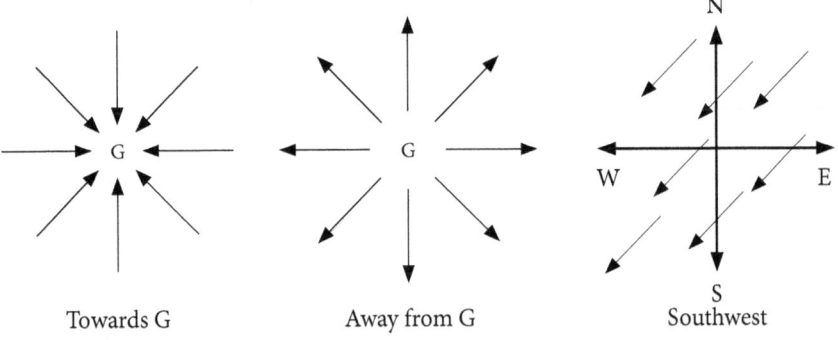

FIG. 5.6 The representation of direction in different frames of reference.

the right-most diagram of Fig. 5.6 have to be described as pointing southwest, regardless of where their tails are. On the other hand, *toward/away from* G does not encode a unique direction (that is, not independently of the location of the Figure). This is another consequence of the polar FoR: all the arrows in the diagram on the left of Fig. 5.6 point toward the Ground G, and all the arrows in the central diagram point away from G.

Clarifying the conditions under which multiple direction specifications encode the same direction presupposes a formal way to capture relationships across directions. The tool used in geometry for this purpose is the 'vector' notion. The meaning of direction expressions in English can be modeled with vectors following the replacement rules in (17):

(17) **Direction vector**: Let D be an expression of the orientation or direction of motion of Figure F with respect to Ground G during the time interval I.[7] Then D denotes the direction vector V, defined as follows:

  a. If D contains *away from*, the location of G defines the tail of V and the location of F during I defines the head of V.
  b. If D contains *toward(s)*, the location of G defines the head of V and the location of F during I defines the tail of V.
  c. If D contains an expression that denotes a non-specific place P in some absolute or relative FoR (such as *up, down, north, left*), then P defines the head of V and the location of F defines the tail of V.

(17c) assumes that specifications in absolute or relative FoRs denote 'non-specific' places, in the sense that it is possible to refer to e.g. any place north of F with *north of F*.[8] The magnitude of V is assumed irrelevant in (17). All other properties of vectors are exactly as defined in geometry. The rules in (17) are specific to the direction terms of English, but it is assumed here that the denotation of direction terms in other languages can be modeled by vectors in an analogous fashion. Given (17), the UVC can be stated as follows:

(18) **Unique vector constraint (UVC)**: all direction vectors denoted in a single simple clause referring to a single continuous motion event must be collinear and of the same polarity. They are interpreted as holding for the entire motion event.

---

[7] G denotes a direction ground in the sense of Jackendoff's (1983: 164) characterization quoted above; the function of G in direction specifications should not be confused with the function of bounded-path and route grounds.

[8] Technically speaking, expressions such as *north of F* seem to quantify existentially over places, so *north of F* does not refer to *all* places north of F but to *some* place north of F. In this respect, the term 'non-specific' may be misleading.

The formulation in (18) is restricted in scope in several non-trivial ways that are discussed below. Point (18) formalizes the intuition that multiple direction specifications are acceptable in single simple clauses as long as they refer to 'the same direction'. The relevant sense of 'sameness' is spelled out geometrically in terms of collinearity and polarity—less technically speaking, two direction specifications are specifications of the same direction if the vectors they denote are on a single line[9] and point in the same direction. In the sense that collinear vectors of the same polarity are identical except for their magnitude and their head and tale locations, the magnitude of direction vectors is here assumed to be irrelevant in semantic representations. Principle (18) is equivalent to (18'):

(18') **Unique vector constraint (UVC)**: all direction specifications in a single simple clause referring to a single continuous motion event must denote the same 'unbounded' direction vector, i.e. the same direction vector after abstraction from head and tale coordinates. This direction vector is interpreted as holding for the entire motion event.

Based on (17) and (18)/(18'), it should be clear that the AUC and the UVC overlap in scope, but neither principle captures all the data accounted for by the other. The AUC cannot explain why (14a) and (14b) cannot refer to the same scenario. This is accounted for by the UVC. The same point is illustrated in (19).

(19) a. Sally walked north away from her house.
b. Sally walked away from her house and then north.

Example (19a) is acceptable just in case the vector encoded by *north* and the vector encoded by *away from her house* (given Sally's location during any time for which the assertion is made) point in the same direction, whereas in (19b), this is not the case. In contrast, the AUC explains the anomaly of (20), which is not accounted for by the UVC:

(20) a. *Sally walked toward the mountain toward her house.
b. ?Sally walked north toward her house.

Both principles can account for the data in (13) above: (13a) and (13b) are anomalous under either principle (with regard to the AUC absolutely and with regard to the UVC as descriptions of Fig. 5.4), while (13c) and (13d) are sanctioned by both principles. Both principles may be invoked in explaining

---

[9] Geometrically, collinearity obtains also across vectors that are on different but parallel lines. This is sufficient to characterize identity of directions in absolute and relative FoRs, but not in the case of *toward* and *away from*.

the data in (10)–(11) above. Examples (11c) and (11d) are again illformed under either constraint, but only the UVC explains why (11a) and (11b) are acceptable as descriptions of certain other scenarios, but not as descriptions of the scenario in Fig. 5.3. Consider (11a): (11a) encodes a single direction vector, specified by *up*. The UVC requires that this vector describes the direction of the entire motion event. This is true just in case motion *over the top* is compatible with motion *up*, i.e. if the top is a slanted (or even vertical) surface, which is not the case in ECOM C6. (Some native speakers of English find constructions similar to *up over the top* applicable even in scenarios where *up* and *over the top* refer to distinct segments of the motion event; such apparent counterexamples to the UVC are addressed in the following section.)

Principle (18)/(18′) introduces three restrictions that require discussion, namely, the condition of structural simplicity, the condition of unique event reference, and the restriction to the syntactic level of the clause. The simplicity condition in (18)/(18′) serves in particular to exclude cases of coordination that can be analyzed as involving gapping, as discussed in sect. 2 with respect to the AUC and in this section with respect to the UVC. Under this analysis, (20a), for instance, is treated as a gapping construction.

(21)   a. Sally climbed up the hill (in the morning) and down again (in the afternoon).
       b. Sally climbed up the hill (in the morning) and climbed down again (in the afternoon).

Like the multi-verb-phrase or multi-clause construction (21b), (21a) refers to the ascension and the descension parts as separate events, as shown by the compatibility of the two adjuncts in (21a) with distinct time adverbials. Therefore, an analysis of *up the hill and down again* in (21a) as a single complex adjunct seems inadequate.

The restriction of the UVC to clauses referring to single continuous events is trivially motivated with respect to iterative or habitual examples, such as those in (22):

(22)   a. (Yesterday,) Sally went to Amsterdam twice.
       b. (Last winter,) Sally skied down the hill every day.
       c. Sally went back and forth between Nijmegen and Amsterdam all week.
       d. Sally commuted between Nijmegen and Amsterdam all summer.

Modifiers that change the frequency or specificity of the event reference obviously do not affect the path of any single instance of the motion event, they merely 'multiply' it. Yet, over the larger interval during which the multiple instances of the event are understood to occur, the Figure must be understood to return to the source of each individual motion event once or multiple times, and hence an entailment of path reversal arises.

Finally, restricting the UVC to the clause level is probably an oversimplification. For example, (21b) may well be analyzed as a coordination of two verb phrases rather than as a biclausal gapping construction. Consider also the following data from the Kwa language Ewe, of Ghana and Togo. In Ewe, reversal of the motion vector may be expressed by a serial verb, *gbɔ* 'come back'. With the main verb *de* 'reach' (but not with *yi* 'go'), this yields a simple serial verb construction that covers the entire trajectory from source back to source, as in (23):

(23) É-de gbɔ.
3.SG-reach come.back
'He went and returned.'

However, as soon as source and goal are overtly specified, a more complex construction has to be used that according to some native speakers requires the connective *hé*:

(24) É-yi Amsterdam tsó Nijmegen (hé-)trɔ́ gbɔ.
3.SG-go Amsterdam from Nijmegen (CON-)turn come.back
'He went to Amsterdam from Nijmegen and came back.'

This suggests that even though the serial verb construction in (23) entails a reversal in the motion vector, the only part of the trajectory that can be mapped onto clause-level syntax, and thus becomes accessible to direction specifications, has to conform to a single vector.[10] Now, interestingly, the construction in (24) is arguably a single clause, as the two parts cannot be negated independently. However, it is not obvious that it represents a single continuous event. Example (25), in which the two parts carry separate time-positional adverbials, is considered fine by some (but again not all) native speakers:

(25) ?É-yi Amsterdam tsó Nijmegen etsɔ (hé-)trɔ́
3.SG-go Amsterdam from Nijmegen yesterday (CON-)turn
gbɔ égbe.
come.back today
'He went to Amsterdam from Nijmegen yesterday and came back today.'

---

[10] The fact that in (24) a more complex construction is needed than in (21) is a consequence of the AUC as much as it is a consequence of the UVC: in (24), two goals are specified. But this does not affect the point that (23) does not violate the UVC because it entails only direction change but no actual direction vectors, while (24) which does entail both a direction vector (away from Nijmegen during some initial subinterval) and direction change employs a more complex

The right generalization is probably that the UVC holds for whatever is an expression of a single continuous motion event in the language. Whether a single continuous motion event is expressed by a clause, a verb phrase, a particular kind of serial verb construction, or something yet different, may depend on the particular language.[11]

The way the 'vector' notion is introduced in (17) to capture the meaning of direction expressions in language is largely identical to the way this notion is used in the Vector Grammar framework of spatial semantics proposed by O'Keefe (1990, 1996, Ch. 4; see also Zwarts, Ch. 3). However, Vector Grammar holds that *all* spatial relations are linguistically represented in a vector format, and that therefore the vector format is superior in linguistic analyses and/or analyses of the workings of the language–cognition interface to other representations of spatial meanings. No commitment is made to this proposal here. In fact, the way vectors are used here to model directional meanings permits a one-to-one mapping from the representations of direction assumed in Jackendoff (1983) into vectors. However, a verification of the UVC would prove that relations across directions matter in the semantics of natural languages. The 'vector' notion is a useful tool to model such relations across directions. The validity of the UVC would thus also provide evidence that vectors are useful tools in semantic analyses. To the extent that Vector Grammar maintains that vectors are a useful tool for linguistic analysis, the validity of the UVC would support this claim in one confined domain of linguistic semantics, namely the semantics of direction expressions (which is the domain where the mapping of spatial meanings into vectors proceeds in the intuitively most straightforward fashion anyway).

The following two sections address two types of apparent counterexamples to the claim that single simple clauses encode no more than a single unique direction vector. The first type is constituted by utterances that implicate direction changes but do not entail them. The second type contains path shape expressions such as the verbs *circle* and *zigzag* and the preposition *around*.

4. The Role of Pragmatics in Direction Coding

It has been argued with respect to the scenario depicted in Fig. 5.4 that the curvature of the extensional trajectory that the motion description refers to has no direct impact on the codability of the motion event in single simple clauses.

construction. This is in line with the UVC (and the AUC) because (24) represents the scenario in terms of two distinct events (witness (25)).

[11] Ultimately, all three constraints in (18)/(18′) may flow from a single one which restricts principles such as the UVC and the AUC to single continuous events.

The UVC affects only the direction *information* that is actually linguistically coded or entailed. However, there must of course be principled restrictions on the possible scenarios a given description can appropriately refer to. These restrictions partly stem from the semantics of the motion event descriptions (which must be truth-conditionally compatible with the scenarios referred to) and partly from pragmatic inference mechanisms of information enrichment in context, as described in particular by Grice (1975), Levinson (2000), and Sperber and Wilson (1986). Consider, for instance, (26) as descriptions of the scene in Fig. 5.7:

(26) a. ?The ant crawled up across the table.
b. ?The ant crawled up over the table.
c. ?The ant crawled up the cloth across/over the table.

For many native speakers of English, (26a–b) are anomalous, at least in reference to Fig. 5.7. Some say that (26c) is clearly better, if still slightly odd. For some native speakers, however, (26a–c) are all perfectly fine, both in reference to Fig. 5.7 and otherwise.[12]

Notice that only one of the two Ground-denoting adjuncts in each of the utterances in (26) encodes a direction vector, namely, *up*. *Over* and *across* merely specify that a line or a surface saliently dividing space in two surrounding regions is traversed from one side to the other, while they do not specify a horizontal or vertical orientation of the line or surface. Therefore, *up across* in (26a) and *up over* in (26b) arguably do not *entail* a change in the direction vector. The utterances in (26) merely trigger *implicatures* of such a change, based on general knowledge about the design and canonical orientation of tables. Indeed, (27) shows instances of *up across* and *up over* without changes in direction that seem at least equally acceptable:

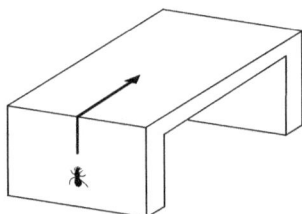

FIG. 5.7 Implicated path curvature: *The ant crawled up over the table.*

---

[12] Whether this variation in native speaker intuitions is dialectal remains to be seen.

(27) a. The ant crawled up (the wall) across the picture.
b. The ant crawled up (the wall) over the picture.

If it is true that *up/down* in combination with *over/across* do not specify a vector change out of context, this might explain why they are apparently more widely acceptable with the cloth scenario in (26c): the cloth can be construed as a continuous surface, abstracting from the horizontal orientation of the table top. Under the analysis presented here, then, the utterances in (26) specify only a single direction vector, denoted by *up*. The path segment denoted by *over/across the table* is oriented horizontally if the table is placed canonically, but that is a matter of world knowledge, not a matter of semantics. The table could be tilted such that its surface would be oriented vertically; (26) would be perfectly fine then.

The question is, if (26) is not at odds with the UVC, why then is it that most native speakers nevertheless find (26) problematic? If it is true that (26) does not specify the orientation of the table, then it gives fairly little information about the actual course of the ant's motion. So how can (26) nevertheless be a useful utterance in an appropriate situation? Because in an appropriate context, a knowledgeable listener will enrich the semantic information of (26) by contextual information and world knowledge. In particular, (s)he will apply Grice's (1975) second maxim of quantity, 'Do not make your contribution more informative than is required', to (26), and infer the most stereotypical construal of (26) that is in line with what (s)he knows or assumes about the situation. Thus, (s)he will infer that the table is indeed oriented horizontally, as tables canonically are, and that the scenario described in (26) accordingly involves a change in direction, as depicted in Fig. 5.7. This inference is a generalized conversational implicature: it is defeasible, but it is none the less the most natural interpretation. Now, what those native speakers who reject (26) apparently object to is that they do not consider it an appropriate description of that most natural scenario that pragmatics calls for. If the scenario in (26) indeed involves a change in direction, then that scenario should be described by a biclausal construction in accordance with the UVC. In short, speakers who reject (26) do so because to them it suggests that the table is oriented vertically. Utterances such as (26) are not *structurally* or *semantically* anomalous according to the UVC, they are *pragmatically* anomalous.

One could advance the argument further and claim that the utterances in (26) are truth-conditionally *false* as descriptions of the scenario depicted in Fig. 5.7. This is because assuming the UVC is correct, these utterances *entail* that the ant moves up *while* it moves over/across the table, and that the table therefore must be oriented vertically, which is not the case in Fig. 5.7. Undoubtedly, this is why most speakers reject (26) as descriptions of Fig. 5.7 (and the same applies to *up over the top* in (11a) above in reference to Fig. 5.3). But speakers who consider (26) fine as descriptions of Fig. 5.7 perhaps compute the entailments and implicatures of (26) in a different order. They first consider the fact that there are *some* scenarios which are truthfully described by the utterances in (26), such as those suggested in (27), and then decide that some feature of Fig. 5.7, such as the presence of the cloth, might license (26) as descriptions of this scenario as well.

5. Path Shape

Prepositions such as *around* and *along* and verbs such as *circle, oscillate, weave,* and *zigzag* encode what may be called **path shapes**:[13]

(28) a. Sally walked around the corner.
 b. The ice skater circled around the monument.
 c. The particle oscillated between the anode and the cathode.
 d. The drunkard weaved along the road.
 e. The dot zigzagged across the screen.

Figures moving along paths of non-linear shapes undergo change of direction. So how can utterances like those in (28) be reconciled with the UVC? Utterances that contain expressions of non-linear path shape do not necessarily violate the UVC because path shape expressions do not by themselves entail directions. Path shapes do not determine directions, they merely determine the change (with non-linear shapes) or preservation (with linear shapes) of directions. Consider Fig. 5.8. The three trajectories have identical shapes, but the direction of the Figure at each point along the first trajectory is the exact opposite of what it is at the corresponding point of the second trajectory, and the relationship between the direction of the Figure at each point along the third trajectory and the directions at the corresponding points of the first two trajectories varies from point to point. Therefore, a description of these trajectories merely in terms of path shape does not entail any direction vector. The UVC concerns only direction specifications. If a description contains such a specification, and in addition entails change of direction, then it becomes subject to the UVC.[14] But if it entails only change of direction, but does not actu-

---

[13] See van der Zee (2000) for a detailed treatment of the semantics of such expressions.
[14] Actually, not all combinations of path shape and direction specifications require multiclausal encoding. Consider e.g. *The river meandered north*. In such cases, a distinction needs to

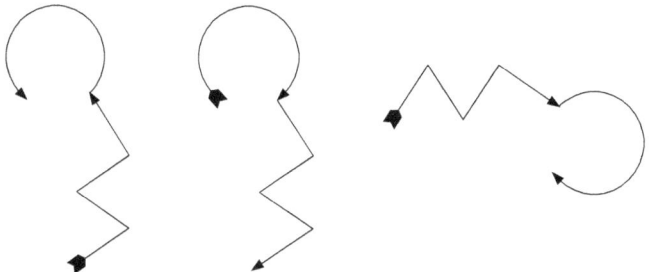

FIG. 5.8 Path shape underdetermines direction.

ally specify a direction vector, then it is not affected by the UVC. The UVC requires by no means *all* motion events that can be encoded in single simple clauses to proceed along straight lines (as far as entailments go!), it merely imposes this requirement on motion event descriptions that specify directions. Consider (29)–(31):

(29) Sally went north.
(30) Sally went around the corner.
(31) *Sally went around the corner north.

Example (29) specifies a direction, but no change. (30) entails change of direction, but does not entail a direction vector as defined in sect. 3. But (31) encodes a direction and in addition entails that at some point during the event, that direction changes. This is ruled out by the UVC.

It is possible, however, to obtain both an entailment of direction and one of direction change in a single simple clause that contains only specifications of bounded path and path shape:[15]

(32) Sally went around the corner to the kiosk.

Example (32) entails that the Figure moved toward the kiosk during appropriately small terminal subintervals. But because of the non-linear path shape denoted by *around*, (32) also entails that Sally was not moving toward the kiosk throughout the event described by (32). This suggests that the UVC concerns

---

be made between **local curvature** (described by *meander*) and **global curvature** (see van der Zee, 2000, on this distinction). Global curvature is not specified in the example. The example is fine in case the global shape of the trajectory is straight and *north* only refers to this global trajectory. On the other hand, the example is anomalous in case the global curvature is e.g. circular.

[15] This was pointed out to me by S. Meira with respect to examples similar to (32).

only coded direction information, but not direction entailments in utterances that do not contain direction expressions.[16]

Conclusions

It has been argued in this chapter that even though there is striking variation in the coding of complex motion events across languages, far greater in fact than previous typological and theoretical studies have suggested, there are also several apparently cross-linguistically and possibly universally valid constraints on how complex motion events are broken down when mapped onto units of linguistic code. Particular attention has been paid to one such constraint, termed the unique vector constraint (UVC), which rules out multiple specifications of direction information in single clauses (or possibly verb phrases), so long as these do not refer to the same direction vector (and the clause or verb phrase codes a single contiguous event). As a model for this constraint (and also in order to delimit the scope of both principles with respect to each other), the argument uniqueness constraint (AUC) has been discussed. The AUC requires each expression of a semantic argument expressed in a single simple clause to be assigned a unique argument role. It has been shown that the AUC also affects the coding of motion path roles such as source and goal. Like the AUC, the UVC is neither a purely syntactic nor a purely semantic or conceptual principle. It is rather a semiotic principle that restricts the coding possibilities at the syntax–semantics interface.

---

[16] It seems plausible that the amount of path shape information that may be coded in single simple clauses is constrained in a similar way to the amount of direction information. Future research will have to show how a constraint on the codability of path shape may be formulated. However, it does not seem possible to derive the UVC and a possible constraint on path shape coding from a single principle. For example, the anomaly of (13a–b) cannot in any straightforward way be accounted for in terms of path shape, because it seems possible to approximately describe the scenario in Fig. 5.4 in a single simple clause in terms of path shape, saying something like *The Figure moved on an L-shaped path from A via B to C*.

6

Defining Spatial Relations: Reconciling Axis and Vector Representations

LAURA CARLSON, TERRY REGIER,
and ERIC COVEY

Abstract

This chapter explores how spatial terms such as *above* are mapped onto spatial regions, with a particular focus on two types of representation: reference frames and spatial templates. The purpose is to argue that the underlying structure of these representations is different. Specifically, reference frames are dependent upon an axial system, whereas spatial templates can be best characterized as a vector representation. In support of this idea, the Attentional Vector Sum model (Regier and Carlson, 2001) is presented. In this model, the direction indicated by a spatial relation is defined as a sum over a population of vectors that are weighted by attention. Conceptualizing spatial templates as a vector representation rather than as an axial representation successfully accommodates a number of factors (orientation, placement relative to the topmost point of an object, and distance) that significantly impact the use of spatial relations. The main conclusion is that both axis and vector representations are necessary for defining spatial relations, with the hierarchical relationship between reference frames and spatial templates offering a means of coordinating these structures.

1. Defining Spatial Relations: Reconciling Axis and Vector Representations

Interpreting spatial relations requires a mapping between language and perception. For example, consider the sentence '*The coin is above the piggy bank.*'

This work was supported in part by NSF grant SBR97-27638 awarded to LAC. Address all correspondence to Laura A. Carlson, Department of Psychology, 118-D Haggar Hall, University of Notre Dame, Notre Dame, IN 46556, email:laura.c.radvansky.2@nd.edu.

and the accompanying picture in Fig. 6.1. The spatial term *above* is used to indicate the location of one object (i.e. the coin; more generally, the **located object**) by referring to a region around a second object (i.e. the piggy bank; more generally, the **reference object**). Comprehension of this sentence requires not only identifying the perceptual referents for the linguistic terms *coin* and *piggy bank*, but also mapping the spatial term *above* onto a particular spatial region. In this chapter we will focus on mapping spatial terms onto spatial regions, focusing on two underlying representations: **reference frames** and **spatial templates**. It has typically been assumed that an axial structure underlies both reference frames (e.g. Carlson-Radvansky and Irwin, 1993, 1994; E. Clark, 1973; Landau and Jackendoff, 1993; Levinson, 1996; Logan and Sadler, 1996; Tversky, 1996), and spatial templates (e.g. Carlson-Radvansky and Logan, 1997; Crawford, Regier, and Huttenlocher, 2000; Hayward and Tarr, 1995; Logan and Sadler, 1996). The purpose of this chapter is to argue that an axial system is not sufficient to characterize spatial term use; rather, a vector representation that best defines the underlying structure of a spatial template (Regier and Carlson, 2001) is also needed. To accomplish this, significant difficulties with conceptualizing the spatial template as an axial structure will be discussed. Next we present the **Attentional Vector Sum model** (ibid.) in which the direction indicated by a spatial relation is defined as a sum over a population of vectors that are weighted by attention. This model is compared with competitor models with different conceptualizations of spatial templates, which in turn are assessed with respect to how well they can accommodate three factors that influence the manner in which the spatial relation is mapped onto space: orientation, placement relative to the outer sides of the reference object, and distance. The attentional vector sum model outperforms the other models, both quantitatively with respect to fitting empirical data, and qualitatively with respect to exhibiting the effects observed in the empirical data. On this basis, we argue that a vector representation underlies spatial templates. Importantly, reference frames and spatial templates can be thought of hier-

FIG. 6.1 Sample picture to accompany the utterance '*The coin is above the piggy bank*'.

archically, in that the orientation indicated by the vector sum of the spatial template can be defined with respect to the axis of a reference frame. In this sense, a spatial template can be considered a parameter of a reference frame, as suggested by Carlson-Radvansky and Logan (1997), thereby coordinating these two types of representations.

2. Parsing Space Using Reference Frames: An Axial System

According to H. Clark (1996), language is a joint activity between a speaker and a listener that is undertaken to accomplish a shared goal. For example, the speaker of the sentence '*The coin is above the piggy bank*.' may have the intention of specifying to the listener the location of the coin, whereas the listener has the intention of finding the coin. The sentence is facilitative because it narrows the space the listener needs to search from an entire room to a region surrounding the (presumed) known location of the piggy bank. Nevertheless, although narrowed, a search domain still exists, and successful mapping of the spatial term onto this space requires an understanding of the underlying representation of this space.

One means of characterizing this space is to define it with respect to a **coordinate system**. According to Hothersall (1995), Descartes's idea of a coordinate system came to mind as he watched a fly buzz around his room, and realized that he could specify its position at any given moment in terms of its distance from the walls and floor or ceiling, with its path of motion characterized by a series of such points. Johnson-Laird (1983) relied on such a coordinate system to show that the interpretation of a spatial term such as *above* could be implemented in a computer program as movement in a specified direction through a two-dimensional array, with locations indicated by the row and column values. More formally, Logan and Sadler (1996) instantiated the coordinate system as a reference frame that consists of a set of orthogonal axes, and that is defined with respect to a set of parameters including **origin, orientation, direction,** and **scale**. The origin is the intersection point of the axes (i.e. where the wall and ceiling meet). Orientation refers to how the axes are aligned (i.e. with respect to the vertical (floor to ceiling) or horizontal (wall to wall) dimensions). Direction refers to the ordering of values along the axes, and is typically defined by the end points of the axes. Finally, scale refers to the units of measure along the axes.

Consistent with many accounts of spatial term use (Carlson-Radvansky and Irwin, 1994; Carlson-Radvansky and Logan, 1997; Garnham, 1989; Herskovits, 1986; Landau and Jackendoff, 1993; Levelt, 1984, 1996; Logan, 1995; Logan and Compton, 1996; Logan and Sadler, 1996; Miller and Johnson-Laird, 1976; Talmy, 1983), apprehension requires imposing a reference frame on the reference object, thereby defining the origin; orienting the axes with respect to a particular source of information, thereby defining orientation and direction; and assigning a distance, thereby defining scale. Depending upon how these parameters are set, different types of reference frame can be distinguished (for an excellent review, see Levinson, 1996), and considerable research has examined preferences for these different types across various manipulations and tasks (e.g. Carlson-Radvansky and Irwin, 1993, 1994; Franklin and Tversky, 1990; Friederici and Levelt, 1990; Levelt, 1982; Schober, 1993). Thus, on both theoretical and empirical grounds, reference frames are necessary for defining spatial relations.

3. Reference Frames and Spatial Templates

Although necessary, an **axis-based system** may not be sufficient to characterize the typical use of spatial relations. As Johnson-Laird (1983: 257) puts it, 'speakers do not normally require that one object lie directly in line with another in order to satisfy the true conditions of a spatial relation. They tolerate a certain amount of vagueness.' Logan and Sadler (1996) formalized this vagueness as a spatial template. Specifically, they presented participants with displays consisting of a reference object (O) that was placed in the middle of an invisible 7 × 7

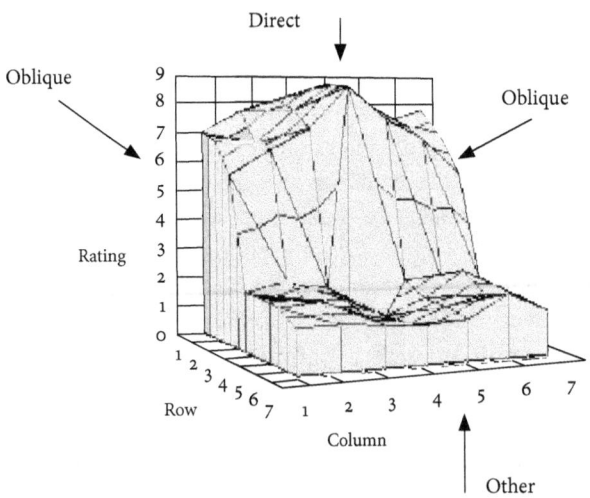

FIG. 6.2 Logan and Sadler (1996, Experiment 2) 'above' spatial template.

grid. Across trials, they placed a located object (X) in the empty cells of the grid. For each placement, participants rated the acceptability of a given spatial relation as a description of the configuration formed by the located and reference objects. The results were represented as a three-dimensional plot, with the rows and columns of the grid defining the x and y axes, and acceptability ratings defining the z axis. The spatial template for the term *above* is shown in Fig. 6.2. Three regions can be identified on the surface of the spatial template: direct, oblique, and other.[1] The direct region corresponds to placements of the located object that are vertically aligned with the reference object; these locations received the highest ratings. The oblique region corresponds to the areas flanking the direct region, locations for which ratings were more intermediate. The other region consists of the remaining regions around the reference object, where ratings were uniformly low.

Importantly, the form of the spatial template shown in Fig. 6.2 is quite general. For example, similar spatial templates are obtained across a range of located and reference objects (cf. letters used by Logan and Sadler, 1996; trees and squares used by Carlson-Radvansky and Logan, 1997; birds and rafts, computers and circles, buildings and offices used by Hayward and Tarr, 1995). Moreover, Carlson-Radvansky and Logan (1997) showed that spatial templates of similar shape were constructed for different types of reference frames. Finally, Logan and Sadler (1996) showed that similar spatial templates are constructed for terms within the same class of relations. For example, the spatial relations *above* and *below* differ only in the end point to which they are assigned on the vertical axis (i.e. the direction parameter of a reference frame); the spatial templates constructed for these relations were similar in shape, although their orientations are flipped across the vertical axis.

4. The Structure of the Spatial Template

These observations suggest a common underlying structure for the spatial template that operates across terms, across objects, and across reference frames. What is this structure? Logan and Sadler (1996) argue that spatial templates are not just plots of data but are psychological representations that assist in the apprehension of **spatial relations**. Specifically, a spatial template is imposed on a reference object, and aligned with the reference frame. This implies that spatial templates are tied to, and perhaps defined by, reference frames.

---

[1] Logan and Sadler (1996) refer to these regions as good, acceptable, and bad; however, following Regier and Carlson (2001), we prefer terminology that refers to the locations of the regions rather than to the acceptability ratings associated with them.

For example, consider the ABOVE spatial template shown in Fig. 6.2. Alignment of the spatial template with the reference frame would map the direct region onto locations along the vertical axis of the reference frame, and would map the oblique and other regions to locations off the axis. This distinction between on-axis and off-axis locations has been an important feature in some accounts of spatial language use (e.g. Crawford, Regier, and Huttenlocher, 2000; Hayward and Tarr, 1995; Munnich and Landau, 1997). For example, Hayward and Tarr argue that spatial relations can be interpreted as categories, and that the highest ratings assigned to the on-axis locations reflect the prototypes of the category. Interestingly, Crawford *et al.* argue that the on-axis locations serve as prototypes for spatial language, but as category boundaries for non-linguistic spatial memory.

One could be even more specific about defining the regions of a spatial template with respect to an axial system. For example, the direct region would correspond to points along the relevant axis (e.g. VERTICAL) at the appropriate end point (e.g. ABOVE), the oblique region would correspond to off-axis locations that are in the same direction from the origin, and the other region would correspond to all locations in the opposite direction. Note, however, that such a definition is merely descriptive, and suffers from potential circularity: The reason that the direct region receives the highest ratings is because those locations occur on the axis of a reference frame; however, the way to infer the location of the axis of a reference frame is to identify the peak in the spatial template.

One way out of this circularity is to obtain an independent means of assessing where the axes of a reference frame are located with respect to the reference object. For the stimuli that have been used in previous studies of spatial templates, this is easily accomplished because only one placement of the located object that is vertically aligned with the reference object is probed (albeit at varying distances) (i.e. see column 4 in Fig. 6.2). In effect, this reduces the reference and located objects to points, with the axis of the reference frame running through these points. This approach is consistent with the idea that use of a spatial relation requires schematization of the reference and located objects, a process by which particular details of the objects are discarded or abstracted over (Herskovits, 1986; Landau and Jackendoff, 1993; Talmy, 1983). At a theoretical level, the typical assumption is that the located object is schematized to a point, and the reference object is schematized either to a point (for example, when the distance between the objects is large, Herskovits, 1998), or to its axial structure (Landau and Jackendoff 1993; Talmy, 1983). As such, the axis of the reference frame would either run through the point or coincide with the reference object's axial structure, and the direct region of the spatial template would then be aligned with this axis.

Defining Spatial Relations 117

However, restricting analyzes of spatial relations to cases in which a single column of locations above the reference object is probed may misrepresent typical use. Reference objects have appreciable extent, and are typically larger (and more salient and stable) than located objects (Huttenlocher and Strauss, 1968; Sadalla, Burroughs, and Staplin, 1980; Talmy, 1983). As such, it is possible to place a located object at multiple locations that are vertically aligned with some *part* of the reference object. This makes inferring the location of the axis of the reference frame more difficult. For example, it is not clear whether the axis of the reference frame would then run through all such points; that would require the axis to be quite thick, with all points aligned with some portion of the reference object designated as within the direct region. It is more likely that the axis would run through just part of the object. The problem then becomes specifying the part upon which the axis would be imposed.

It is possible that specification could be based on a geometric characterization of the reference object. For example, Herskovits (1998) argues that when the distance between the reference and located objects is small, the shape of the objects and their relative placement may alter spatial term usage. Similarly, Johnson-Laird (1983) states that the vagueness associated with use of a spatial term is likely to be affected by the sizes and shapes of the objects being related. An alternative possibility is that functional properties of the object may influence how the object is characterized. Evidence in favor of functional effects on the perception of objects (e.g. Lin and Murphy, 1997; Schyns, Goldstone, and Thibaut, 1998) and functional influences on the selection of a reference frame (Carlson-Radvansky and Radvansky, 1996) support this view. Carlson-Radvansky, Covey, and Lattanzi (1999) recently tested this idea by contrasting a geometric characterization of the reference object against a characterization in which the identity of the object and its function were important for determining the best use of the spatial term. Within the geometric account, the reference object was schematized to a bounding box, and its center-of-mass was calculated (Regier, 1996, 1997); the assumption was that the relevant axis of the reference frame would be imposed at this location. In contrast, within a functional account, the part of an object most critical for its function is considered particularly salient (e.g. Lin and Murphy, 1997); as such, one might expect the reference frame to be imposed over the center of mass of this functional part.

The **geometric and functional views** were distinguished by showing subjects pictures of objects for which the functional part was offset from its center of mass, as illustrated by the toothbrush in Fig. 6.3. Participants were then handed a picture of a located object (e.g. a tube of toothpaste), and asked to place it in a specified relation around the reference object. Placements were recorded as deviations from two lines, one running through the **center-of-mass** and one

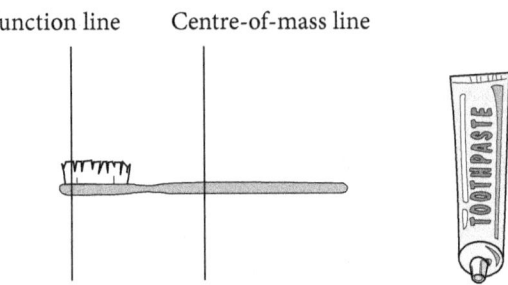

Place the tube of toothpaste above the toothbrush.

FIG. 6.3 Illustration of offsetting the functional part (bristles of the toothbrush) from its center-of-mass. Placements were measured relative to the function and center-of-mass lines.

running through the **functional part** of the reference object. The results were that placements were significantly biased in the direction of the functional part, suggesting that a strict geometric characterization of the reference object is not sufficient to predict how a spatial relation is mapped onto the space surrounding a reference object. More importantly, for current purposes, the mean deviation corresponded to a point that was 59 percent of the distance from the center-of-mass line in the direction of the functional line. Interpretation of this point is not straightforward. Does it reflect the linear combination of two reference frame axes, one imposed on the center-of-mass, and one imposed on the functional part? Or does it reflect the placement of a composite reference frame whose origin is determined by a linear combination of differentially weighted geometric and functional sources (see Carlson, 2000 for a more detailed discussion of these possibilities; see also Coventry and Prat-Sala, 1998; Garrod, Ferrier, and Campbell, 1999). Thus, a significant challenge for defining spatial templates with respect to an axial structure is independently determining where the axis of a reference frame is imposed, and how its placement emerges as a function of geometric and functional influences.

5. The Attentional Vector Sum Model

An alternative approach is to conceptualize spatial templates in a manner that is independent of how the reference frame is established. If spatial templates are independent representations, then it is possible that their underlying structure is different from the axial structure that underlies a reference frame. One possibility is to use vectors rather than axes to define the spatial template. Regier and Carlson (2001) present the **Attentional Vector Sum (AVS)** model in which a spatial template is conceptualized as a population of vectors that are differentially weighted by attention. The architecture of the model is informed by two independent observations: First, human apprehension of spatial relations involves **attention**. For example, Logan (1994) found that visual search for a target in a field of distractors is slow when targets differ from distractors in the spatial relation between their elements, implicating a role for attention in the computation of spatial relations. Second, in several neural subsystems, **overall direction** is represented as the **vector sum** of a set of constituent directions. For example, Georgopoulos, Schwartz, and Kettner (1986) examined a population of orientation-tuned cells in the area of monkey motor cortex that represented the animal's arm. Each cell was broadly tuned, with a preferred direction, such that the cell would respond maximally when the monkey's intended arm movement was in the preferred direction. The direction of motion of the arm was accurately predicted by a vector sum over the population of cells. More recently, Wilson and Kim (1994) found a vector sum representation in human perception of motion direction. This suggests that this type of representation may be widely used.

The AVS model brings together these two observations concerning attention and vector sum. The important features of the AVS model are illustrated in Fig. 6.4. In the presentation of the model, we focus on the term *above*, while noting that AVS generalizes to other projective relations (e.g. *below, left, right*). An **attentional beam** is focused on the reference object at the point that is vertically aligned with the closest part of the located object (see Fig. 6.4, panel (a)). Consequently, parts of the reference object nearest to the located object are maximally attended; more distant parts are attended less. This yields a distribution of attention across the reference object. In addition, vectors are defined that are rooted at positions across the reference object, and that point to the located object. This results in a population of vectors (panel (b)). The vectors are weighted by the amount of attention being paid at the location of their roots (panel (c)). The model then computes the sum over this population of weighted vectors, yielding an orientation (panel (d)) that can be compared to upright vertical (panel (e)). Note that while AVS thus requires a reference frame axis with which to compare the orientation that it outputs, it does not require specification of the placement of the axis.

The orientation that is output by AVS is then mapped onto an acceptability rating that indicates how well a given spatial term describes the relationship between a located object and a reference object. Perfect alignment with

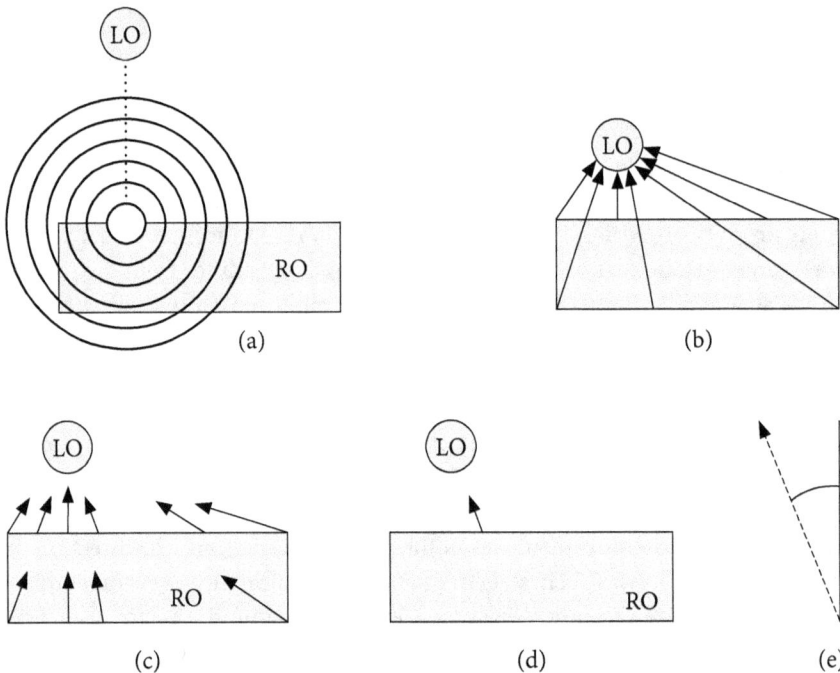

FIG. 6.4 The attentional vector-sum (AVS) model. Panel (a) illustrates the attentional field, focused on the reference object at a point nearest the located object. Different parts of the reference object receive different amounts of attention. Panel (b) illustrates the vectors rooted at each point of the reference object, pointing toward the located object. Panel (c) illustrates the **attentionally weighted vectors**. Panel (d) illustrates the **direction of the attentionally weighted vector sum**. Panel (e) illustrates the orientation of the vector sum relative to vertical upright. RO = reference object; LO = located object. Copyright © 2001 by American Psychological Association, Regier, T. and Carlson, L. A. (2001). Grounding spatial language in perception: An empirical and computational investigation. Journal of Experimental Psychology: General. Vol 130(2): 116–39.273–98. Adapted with permission.

upright vertical would thus receive a perfect rating, with ratings decreasing as deviation from upright increased.

5.1. Competing Conceptualizations

The AVS model isn't the only possible way to conceptualize a spatial template. One alternative would be to define the space with respect to a bounding box that encloses the reference object. This idea is instantiated as the **Bounding**

Box (BB) model in which the applicability of the spatial relation *above* is defined as a region of space bounded by three straight lines: two vertical lines, each extending from the left or right edges of the landmark, and a horizontal line that runs along the topmost surface of the reference object (the so-called grazing line, see below).

Whereas the BB model is expressed in Cartesian coordinates, a second alternative is to define spatial templates with respect to polar coordinates. For example, Regier (1996, 1997; see also Gapp, 1995) proposed that space around the reference object can be defined with respect to two angular features: proximal orientation and center-of-mass orientation. Proximal orientation is the angular deviation relative to upright vertical of a line connecting the closest two points of the located and reference objects. Center-of-mass orientation is the angular deviation, relative to upright vertical, of a line connecting the centers-of-mass of the located and reference objects. This idea is instantiated in the **Proximal and Center-of-Mass (PC) model**, in which the acceptability of *above* for a given located object's placement around a reference object is a product of its proximal and center-of-mass orientations.

Finally, a third alternative is **a hybrid model** that combines the height component of the BB model with the angular features of the PC model, yielding **the PC-BB model**.

5.2. Evaluating the AVS Model

Regier and Carlson (2001) present empirical data from human participants, model fits, and model predicted data from seven experiments that were designed to test among the AVS, BB, PC, and PC-BB models. Initially, all the models were trained on the *above* spatial template data from Logan and Sadler (1996), and the parameters within the models were fixed. The models were then 'shown' displays containing located and reference objects, and each model produced an acceptability rating corresponding to how well a given spatial term described the relationship between the two objects. The located object was schematized to a point, and a reference object (of many different shapes) was used that had two-dimensional extent. Use of such reference objects forces the models to confront the possibility of multiple placements along the top surface of the reference object, a problem that an axial-based conceptualization of a spatial template has difficulty addressing. The models were then evaluated both quantitatively, in terms of how well they fit the empirical data from each experiment, and qualitatively, in terms of whether the model-predicted data exhibited the same kinds of effects as the empirical data. In terms of specific tests, the performance of AVS and its competitors was evaluated on three factors that influence how the spatial relation is mapped onto space around the reference object: proximal and center-of-mass orientation, the grazing line, and distance.

5.2.1. Proximal and Center-of-Mass Orientation

If **proximal and center-of-mass orientations** are important for defining spatial relations (Regier, 1996, 1997), then variations in the shape of the reference object should influence acceptability judgements (Herskovits, 1998; Johnson-Laird, 1983). This is because as shape changes, proximal and center-of-mass orientations change, as illustrated in Fig. 6.5. In panel (a), both the proximal orientation (dashed line) and center-of-mass orientation (solid line) are perfectly aligned with vertical upright. In panel (b), the proximal orientation is still aligned with upright vertical but the center-of-mass orientation deviates. In panel (c), the center-of-mass orientation retains the value it had in (b), but the proximal orientation is more deviant.

Across four experiments, Regier and Carlson (2001) demonstrated that human observers are sensitive to center-of-mass and proximal orientation, such that acceptability decreased as deviation from upright vertical increased.

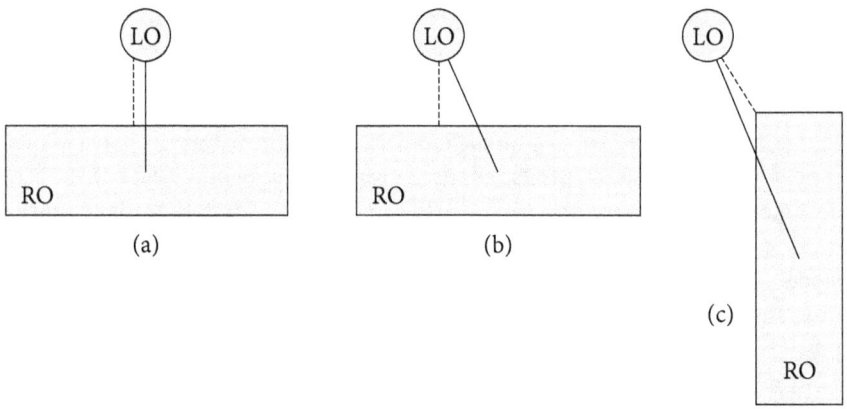

FIG. 6.5 In panel (a), both the proximal and the center-of-mass orientations are perfectly aligned with upright vertical; in panel (b), the proximal orientation is still aligned with upright vertical but the center-of-mass orientation deviates somewhat; in panel (c), the center-of-mass orientation retains the value it had in (b), but the proximal orientation is more deviant. Solid line = center-of-mass orientation; Dashed line = proximal orientation; RO = reference object; LO = located object. Copyright © 2001 by American Psychological Association, Regier, T. and Carlson, L. A. (2001). Grounding spatial language in perception: An empirical and computational investigation. Journal of Experimental Psychology: General. Vol 130(2): 116–39.273–98. Adapted with permission.

Defining Spatial Relations 123

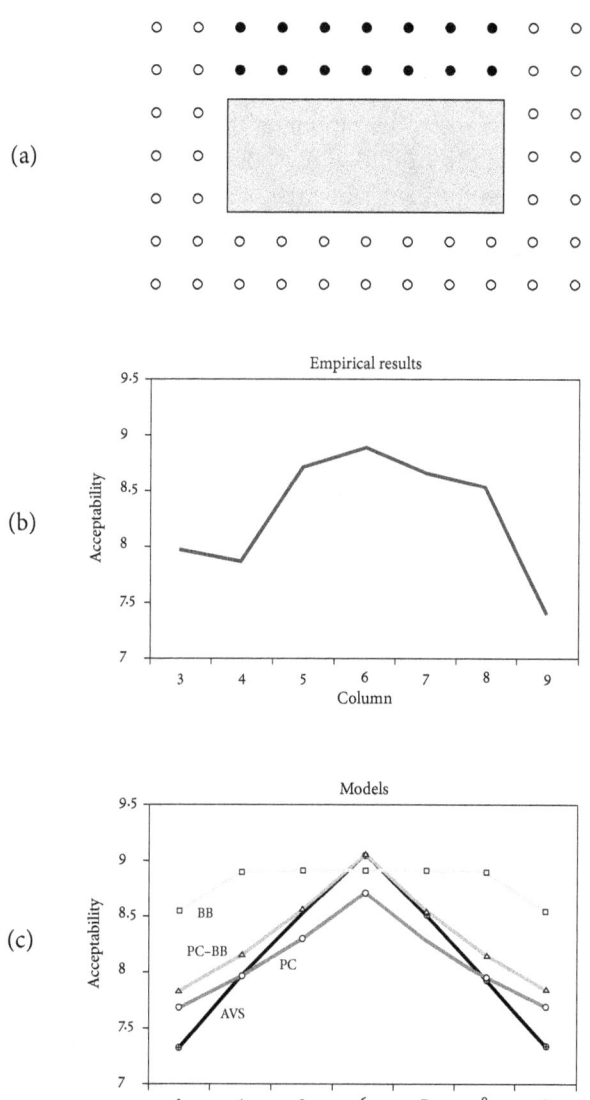

FIG. 6.6 Panel (a) shows critical placements around a rectangular reference object. Panel (b) shows empirically obtained ratings for the critical points (filled circles) in panel (a). Panel (c) shows the model-predicted ratings. Rating scale ranged from 0 (not acceptable) to 9 (perfectly acceptable). Copyright © 2001 by American Psychological Association, Regier, T. and Carlson, L. A. (2001). Grounding spatial language in perception: An empirical and computational investigation. Journal of Experimental Psychology: General. Vol 130(2): 116–39.273–98. Adapted with permission.

For example, in one experiment, subjects were shown a rectangular reference object and a point-like located object that was placed, across trials, in various locations around the reference object. The task was to rate the acceptability of the spatial relation *above* as a description of the configuration between the objects on a scale ranging from 0 to 9. As shown in Fig. 6.6a, critical placements (filled dots) varied the center-of-mass orientation while holding constant the proximal orientation. Figure 6.6b shows the empirical data; the decrease in acceptability as a function of center-of-mass orientation is evident by the drop-off. Figure 6.6c shows the ratings predicted by the various models. AVS, PC, and PC-BB all show a drop off as a function of orientation; BB exhibited this effect only very weakly. In a comparable experiment that held center-of-mass constant while manipulating proximal orientation, both human observers and AVS, PC, and PC-BB exhibited significant proximal orientation effects; BB did not exhibit this effect. These failures to account adequately for center-of-mass and proximal orientation effects thus rule out the BB model.

Within the PC and PC-BB models, these orientation effects emerge directly from their architecture; as such, the models concretized these effects as mechanisms. Within the AVS model, these effects emerge from the attentional vector sum operation in conjunction with the **attentional field**. Specifically, the attentional field can be narrowed so that only a single point on the reference object receives any attention; this point will be the one that is closest to the located object. In this case, the vector sum reduces to proximal orientation. In contrast, the attentional field can be made infinitely wide, so that there is no drop-off in attention across the reference object. In this case, the vector sum reduces to center-of-mass orientation (see Regier and Carlson, 2001, appendix A). For attentional fields of intermediate widths, AVS produces vector sums that will exhibit both effects. Thus, a conceptualization of a spatial template consistent with AVS can accommodate sensitivity to proximal and center-of-mass orientation.

5.2.2. *The Grazing Line*

The second factor to influence the manner in which the spatial term is applied to space around the reference object was placement of the located object above or below a horizontal line running through the topmost point of the reference object. We refer to this line as the **grazing line**. Figure 6.7 presents data from an experiment that tested for an effect of the grazing line while holding proximal and center-of-mass orientation constant. The filled dots represent various placements of the located object across trials around the triangular reference object. The numbers by the dots are the mean acceptability ratings for *above* as a description of the configuration formed by the located object at that placement and the triangle. The critical points occur on Lines 1 and 2. Within a line, a pair of points can be compared for which the proximal and center-of-mass orientation is the same (as illustrated by the dashed lines extending from the points to the reference object). Importantly, one dot in each pair occurs above the grazing line, and the other dot appears below the grazing line. If spatial acceptability judgements are sensitive to placements relative to a grazing line,

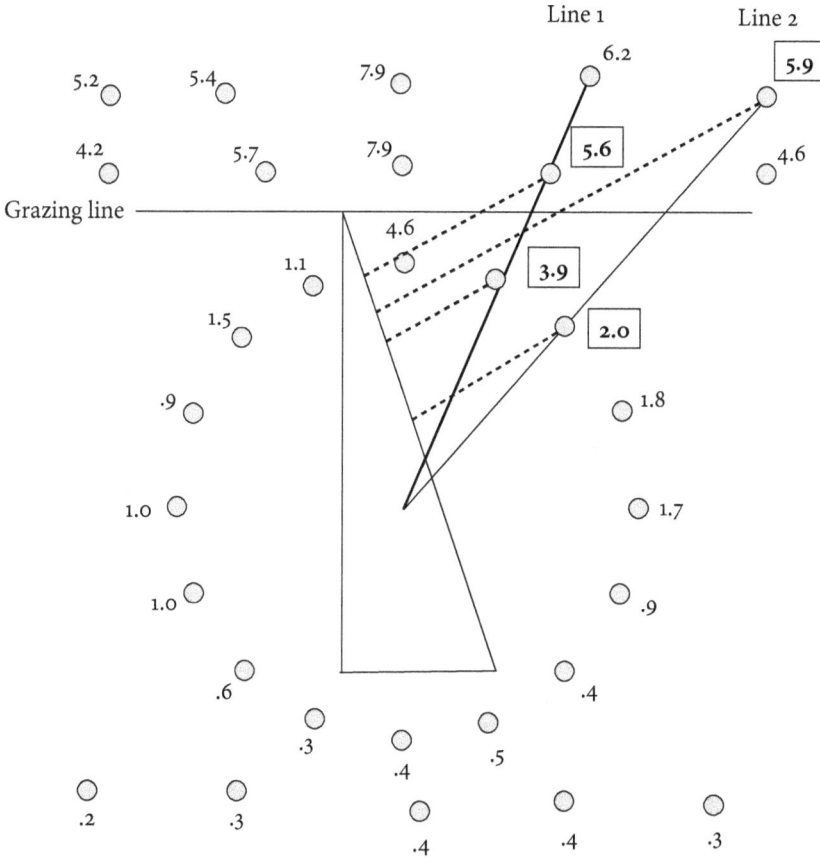

FIG. 6.7 Placements and mean acceptability ratings around the tall triangle reference object. The four critical points on the two radiating lines are outlined and appear in boldface. Solid radiating lines = center-of-mass orientation. Dashed lines = proximal orientation. Copyright © 2001 by American Psychological Association, Regier, T. and Carlson, L. A. (2001). Grounding spatial language in perception: An empirical and computational investigation. Journal of Experimental Psychology: General. Vol 130(2): 116–39.273–98. Adapted with permission.

then there should be a significant difference in ratings between the dots within a pair. This result was obtained for the pairs of dots on both lines.

Importantly, when tested on these points, AVS and PC-BB all demonstrated a qualitative grazing line effect in their predicted ratings, and quantitatively fit the empirical data tightly. This is due to the fact that the grazing line is instantiated in these models as a height function that gates the value yielded by the model. When the placement of the located object is above the grazing line, the height function returns a value of 1; when the placement is below the grazing line and below all points along the top surface of the reference object, the height function returns a value of 0. Points that are below the grazing line but above other parts of the reference object receive intermediate values. In contrast, the PC model fails to pass this test because it has no means of incorporating vertical distance. Note that the BB model also has a height function, it exhibits these effects, and fits the data. However, recall that this model was conceptually ruled out due to its failure to account adequately for proximal and center-of-mass orientation effects. Consequently, the AVS and PC-BB models are the only remaining contenders.

5.2.3. *Distance*

The third factor to influence spatial term use was an influence of distance on sensitivity to centeredness above the reference object. This factor was based on the following intuition. Imagine holding a small marble half an inch above a large book lying on a table, and moving the marble over the surface of the book at a constant half-inch height. At such low heights, use of the spatial term should be somewhat insensitive to the degree to which the marble is close to the center of the book. In effect, the marble is moving across a limitless plane. In contrast, if the marble is much higher above the book, the edges of the book intuitively begin to impact judgements, such that the centeredness of the marble with respect to the book now becomes more important.

This interaction between distance and centeredness is accommodated within AVS by the attentional field. When the located object is very close to the reference object, the attentional field will be narrow, and not much of the reference object, including most especially the edges, will receive appreciable amounts of attention. However, when the located object is far from the reference object, the attentional field is wider and will begin to take in the edges of the reference object; this leads to a greater sensitivity to centeredness. In contrast, the PC-BB model makes the opposite prediction. That is, as distance from the reference object increases, the proximal and center-of-mass orientations become less deviant; thus, the effect of centeredness should diminish with distance.

These predictions were tested using displays in which the placements of the located object were similar to those shown in Fig. 6.6a, except that there was a large separation between the top two rows, such that the upper row was far from the top of the reference object, and the lower row was very close to the top of the reference object. The AVS model predicts greater centeredness effects for the far row than for the close row; in contrast, the PC-BB model predicts greater centeredness effects for the close row than for the far row. Figure 6.8a shows the empirical data. Note the significant interaction between row and distance, with the effect of centeredness larger for the far row (solid line) than for the close row (dashed line). Figures 6.8b and 6.8c show the AVS-predicted and PC-BB-predicted ratings, respectively. The similarity between the empirical data and the AVS ratings, but not the PC-BB ratings, is evident. In addition, AVS provides a better quantitative fit to the data than the PC-BB model.

6. Spatial Templates: Vectors not Axes

Regier and Carlson (2001) clearly demonstrate that a conceptualization of a spatial template as a **vector representation**, as instantiated within AVS, can accommodate the influences of proximal and center-of-mass orientation, the grazing line, and distance. In contrast, these factors cannot all be accommodated within a single competitor model (BB, PC, PC-BB), although individual models do accommodate individual factors. The implication is that spatial templates are best characterized as a vector-based representation.

Note, however, that none of the competitor models have an explicit axial component. Perhaps, then, these models do not adequately represent an axial-based model. To be a viable competitor to AVS, at the minimum, an axial-based model would have to account successfully for the influences of these factors. Let us evaluate whether this is possible. First, consider proximal and center-of-mass orientations. These orientations are measured relative to a reference line; for the spatial term 'above', this reference line is vertical upright. With respect to a reference frame, placements aligned with the reference line (0 deviation) could be considered to fall along the axis of a reference frame, in the direct region. Placements that deviate from upright would fall within the oblique, or other region, depending upon the direction of the placement. However, simply labeling the placements as direct, oblique or other is a mere description; it does not explain why one might expect lower ratings for these off-axis placements as opposed to on-axis placements. The importance of the AVS model is that it can account for these ratings not by describing their placement in a region or relative to the axis *per se*, but by determining the orientation by summing

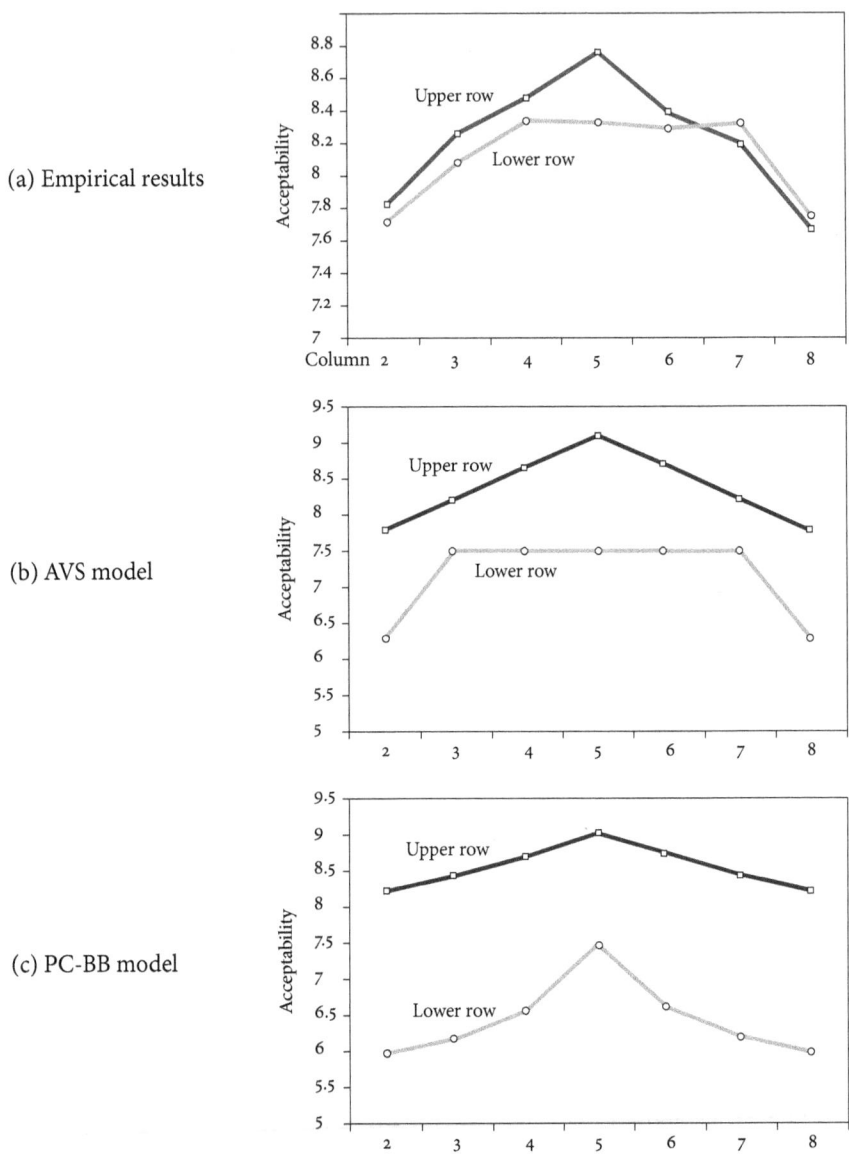

(a) Empirical results

(b) AVS model

(c) PC-BB model

FIG. 6.8 Panel (a) shows empirical data for critical placements above a rectangular reference object (as in Fig. 6.7), with one row close to the reference object and one row further away. Panel (b) shows predicted output for the AVS model. Panel (c) shows predicted output for the PC-BB model. Copyright © 2001 by American Psychological Association, Regier, T. and Carlson, L. A. (2001). Grounding spatial language in perception: An empirical and computational investigation. Journal of Experimental Psychology: General. Vol 130(2): 116–39.273–98. Adapted with permission.

across vectors that are weighted by attention. In effect, the oblique, and other placements receive lower ratings because they are less-attended. Clearly, an axial system that divides space into on-axis versus off-axis locations, or direct, oblique, and other regions is insufficient for characterizing the form of the spatial template. Even defining the three regions of the template with respect to the reference frame falls short of an explanation of the corresponding ratings.

Second, with respect to the grazing line, an **axial-based system** may account for this effect if it is assumed that the origin of the reference frame is imposed at a point along this line (i.e. the apex of the triangle in Fig. 6.7). If so, then one of the horizontal axes of the reference frame could serve as the grazing line. However, this account also requires an understanding of how the axes of a reference frame are imposed on the reference object, an issue that is far from settled. Note, too, that the grazing line is the topmost surface when the term being considered is *above*. When the term *below* is being considered, the grazing line would correspond to the bottom-most surface, and similarly the left and right edges for the terms *left* and *right* respectively. More generally, then, looking across projective terms, the grazing line can be thought of as a bounding box. However, as Regier and Carlson (2001) have shown, a model based on the bounding box alone (the BB model) or in combination with the proximal and center-of-mass orientations (the PC-BB model) is not sufficient to characterize the space around the reference object.

Third, an axial-based system fails to account adequately for the influence of distance. For example, Logan and Compton (1996) failed to find effects of distance when the located object was placed at various distances directly above or below the reference object. This pattern is consistent with work on spatial templates that typically shows no drop in acceptability within the direct region (Carlson-Radvansky and Logan, 1997; Hayward and Tarr, 1995; Logan and Sadler, 1996). However, as shown in Fig. 6.8a, distance effects appeared across the direct region, especially at the centermost location where the axis of the reference frame would presumably be placed.

Thus, an axial-based system is too restricted to account for factors that influence the manner in which the spatial term parses space in off-axis locations, as demonstrated by the failure to account for orientation effects, or in on-axis locations, as demonstrated by the failure to account for distance effects. Moreover, aligning the grazing line with one of the horizontal axes of the reference frame requires an understanding of how the origin is imposed on the reference object. Alternatively, conceptualizing the grazing line more generally as a bounding box will ultimately not be successful. Therefore, spatial templates cannot be defined solely with respect to axial structure; something additional is necessary. Perhaps it would be possible to add properties (e.g. proximity to an edge of the reference object, consideration of spatial or functional characteristics) that would make an axial-based model possible. However, the addition of these components would be post-hoc, motivated by data rather than independently motivated, as are the components of AVS. Moreover, the interaction among the components would have to be specified. Until such a model is developed to a level that it could be tested against the AVS model, the weight of evidence is in favor of AVSs characterization of a spatial template as a vector representation.

7. Relating Reference Frames and Spatial Templates

The argument that reference frames have an axial structure while spatial templates depend on a vector representation does not undermine the relationship between these two representations during the apprehension of spatial relations. Both are necessary representations that have unique processes that govern their activation and use (Logan and Sadler, 1996). Indeed, Carlson-Radvansky and Logan (1997) demonstrated that a spatial template was constructed for each active reference frame, and that the acceptability of using a particular spatial term could be predicted by a linear combination of these differentially weighted spatial templates. As such, Carlson-Radvansky and Logan argue that spatial templates can be considered another parameter of a reference frame. In this sense, the nature of the relationship between these representations is hierarchical. Nevertheless, as a parameter, spatial templates may retain their own structure and underlying representation, much as the scale parameter of a reference frame may have its own underlying metric.

AVS offers a means of hierarchically relating these representations. For example, an axis of a reference frame could serve as the reference line that is used within AVS for the calculation of orientation. Indeed one advantage of the AVS model is that it does not require specification that the reference frame be placed at any particular location, thus subverting the problem addressed earlier of how the origin is defined and the axes are set. Clearly, though, the axes have to be set somewhere, and current work on AVS is focused on how geometric and functional characteristics of the reference object may define where they are set. Conceptually, this is a natural extension of AVS, with a functional influence easily incorporated into the attentional field and vector sum operation. For example, rather than have a single attentional beam that radiates out from one location within the reference object, multiple beams could be used, one focused on the geometric center of mass of the reference object and the other focused at its functional part, with the strength of these beams differentially weighted. Alternatively, a single attentional beam could be used that is weighted toward the functional part, effectively tilting the overall vector sum. The hope is that a successful implementation of the functional influence within AVS will give some insight into the manner in which geometric and functional influences interact to define where the reference frame is imposed on the reference object, and consequently how the alignment of the spatial template with the reference frame is accomplished.

8. Conclusion

We have argued that reference frames and spatial templates are necessary representations that are used during the apprehension of spatial relations (Logan and Sadler, 1996). However, although they operate together, they have different underlying structures, with a reference frame consisting of an axial structure and a spatial template consisting of a vector-based representation. Thus, we argue that both types of representations (axes and vectors) are necessary for characterizing spatial language. The challenge for future work will be a better understanding of the processes that operate to coordinate these representations.

7

Places: Points, Planes, Paths, and Portions

BARBARA TVERSKY

Abstract

This chapter considers the information people use when comprehending and producing directional expressions to describe object location. It concludes that people tend to make use of landmarks, but tend to avoid using explicit specifications of direction and distance. The latter seems to be caused by the difficulty of computing direction and distance, compared to computing landmarks.

1. Ways of Talking About Spatial Locations

SPOUSE 1: *Honey, where's the baby?*
SPOUSE 2: *Here. At 132 Maple. By the sofa. In front of me. Two feet behind you. Go down the hall, go right, then go down until you get to the kitchen. At Joey's house, which is just north of the library and east of the coffee shop.*

How do we tell someone where something is? This simple question turns out to have not so simple answers. The first thing to notice about the answers to the 'where' question is that they are relative to a **reference object** and **frame of reference**. The next thing to notice is that the selection of reference object and reference frame depends on what the information provider, Spouse 2, presupposes about the state of knowledge of the information receiver, Spouse 1. Finally, providing location often entails other information, such as **direction** and **distance**. Let us examine the answers more closely, as they reveal some of the techniques that people use to convey location.

Here. This is the easiest way to convey location, if it works. Like all answers, it depends on common ground between provider and receiver of information. Here, they must share an understanding of where *here* is and what *here* is. *Here* is context-bound. For one thing, it means something different when interlocutors are in close proximity than when they are distant; it may refer to one's person or one's country.

At 132 Maple. Another direct way to convey location is to give an address. At first, an address like 132 Maple Avenue, Highland Park, Illinois or the GPS coordinates for that location might seem like absolute locations, but of course they are relative as well, referring to streets in cities in states or to longitude, latitude, altitude relative to Greenwich. Nevertheless, if the speaker has reason to believe that the listener knows the reference points and frame, then specifying a location with an address can be very effective. What makes addresses so easy is that they refer to points or to regions or volumes that can be conceived of as points (e.g. Talmy, 1983). The baby's at 132 Maple; exactly where in the house or on the lot is not needed.

By the sofa. It is not always practical or appropriate to convey location by means of an address. Fortunately, people have many other ways to do that. Another simple way to convey location is relative to a **landmark**, in this case, the sofa. Like an address, using a landmark to convey location only requires thinking about points, in this case, proximity to a point rather than identity with a point. This way of locating is slightly more complex than an address as both a landmark and a relationship to a landmark are indicated. However, it is not necessary to know or state the exact spatial relationship of the landmark to the object.

In front of me. This way of specifying location also uses a landmark, *me*, and a spatial relationship, specifically, a direction, *in front of*. However, both landmark and spatial relationship are more complex than in the preceding *by the sofa* example. The landmark is regarded as a volume, not as a point, and not as an undifferentiated volume, but as one with intrinsic sides, typically six of them, front, back, left, right, top, and bottom. Direction is specified as a projection from one of the intrinsic sides. Here, it is conceived of in two or three dimensions; it refers to a portion of the space around the landmark, the region projected from the intrinsic side.

Although a portion may seem too large to help specify location, people seem to agree on the most likely places within the portion (e.g. Franklin, Henkel, and Zangas, 1995; Hayward and Tarr, 1995; Logan and Sadler, 1996; Regier, 1996). They even agree on locations that are hedged, such as *mid-left* or *far right* (Franklin *et al.*, 1995). People also condition their expectations about places in portions depending on the context, for example, *near* a wall is understood to be a different distance for ants than for elephants (Morrow and Clark, 1988).

Two feet behind you. As before, this way of specifying location adds an additional element of complexity to the preceding. This expression includes a landmark, a projected direction, and a distance. Not only is adding distance an additional complexity, another feature to keep in mind, but it is also a feature that is vague. People's minds do not contain yardsticks or scales. On the contrary, people's estimates of distance, weight, size, and other quantities are approximate, despite metric terminology (e.g. Leibowitz et al., 1993).

Go down the hall, then go right and go down until you get to the kitchen. This answer is again more complex than the previous ones. Instead of specifying a location, it specifies a procedure for arriving at the location. Such an expression can require all the previous elements (in this case, for brevity, it eliminates some) as well as conceptions of paths and turns. Paths can be specified similarly to portions, above, as directional projections from intrinsic sides. However, they are conceived of as one-dimensional lines. They can also be specified as links between landmarks. Turns are with reference to paths; they are typically specified as projections from intrinsic sides at a specified point and orientation on a path. Such procedures have been termed **route descriptions** (e.g. Perrig and Kintsch, 1985; Taylor and Tversky, 1992*b*). They describe the locations of landmarks relative to the intrinsic sides of an observer changing position and orientation along paths in an environment.

At Joey's house, which is just north of the library and east of the coffee shop. This answer illustrates yet another complex way of specifying location. Instead of prescribing a method of arriving at the location, it sets the scene the target is part of and specifies the location within the scene. This way of specifying location rests on the locations of other landmarks and the directions projected from the surrounding environment. Such scene settings for spatial location have been termed **survey descriptions** (e.g. Perrig and Kintsch, 1985; Taylor and Tversky, 1992*b*). They describe the locations of landmarks relative to one another using the directions of the encompassing environment.

Note the richness of spatial language that is readily invoked simply to say where the baby is. And this is only a hypothetical example; real-life examples usually supply additional surprises. Nevertheless, let us review the types of spatial language that can be used to specify location that these examples illustrate. Points, planes, paths, and portions seem to be the key elements, along with directions and distances. **Points** are typically landmarks, identified by names. They can be intersections, or other features; the essence of points is that they are conceptualized as dimensionless locations. Landmarks in and of themselves can be conceived of as points, but also as one-dimensional paths, two-dimensional planes, or three-dimensional volumes. **Paths** are one-dimensional connectors. They can be identified in several ways: as lines between landmarks, as known roads or links in an environment, or as lines emanating at a direction from a landmark. In everyday speech, directions are given approximately, not in degrees (e.g. Franklin, Henkel, and Zangas, 1995). **Direction** can be expressed relative to the intrinsic sides of the landmark (viewed as a dot or a line or a plane or a volume) or relative to the encompassing environment (viewed as a plane or a volume). **Distance** can be expressed in standard units such as feet and miles. More typically in spontaneous speech, distance seems to be expressed in the approximate units of experience, such as blocks or even time. Route and survey (scene-setting) descriptions use combinations of points, paths, portions, distances, and directions.

Expressing location, then, can invoke considerable complexity and richness of spatial language. What elements seem to be preferred, and why? It is tempting to make a case for simplicity, the less the better. Specifying location is a communication task, typically between two people, but conceivably between the same person at two different times. Communication is easier when there is less to compute in order to produce an utterance and less to compute in order to understand and to remember an utterance. Thus, the emphasis on ease and simplicity takes into account both the cognitive and the social nature of communication. Both provider and receiver are presumably to some degree aware of what is easy for each to produce and comprehend. In actual communication, simplicity does not always prevail; for one thing, communication is fallible, so redundancy diminishes error.

Here, I present evidence from some of our research that shows preferences for using some kinds of information over others in specifying location. First, the conclusions. Using a landmark alone should be ideal, where it is sufficient. Adding direction information is often needed, but computing and comprehending directions adds to the cognitive burdens of both providers and receivers of spatial information. Some directions are easier to compute and comprehend than others, and the easier ones should be preferred. Directions are more easily conveyed approximately, typically as categories plus hedges, than analogically, as, for example, projections from the sides of the body or the cardinal directions rather than degrees. Distance information is problematic for one of the reasons direction information is problematic. People have only vague and approximate knowledge of distances, so distance information can be unreliable. Finally, minimal utterances may be easier to compute and to comprehend, but minimal utterances run the risk of being misunderstood or forgotten. Most effective human communication is redundant.

In the interest of redundancy, I repeat the claim: To express location, where possible, landmarks should be preferred, directions should be avoided, especially those that are hard to compute and comprehend, and distances should be avoided. Direction and distance information will be given when proximity to landmarks is not sufficient, but it will be in hedged categories. The evidence

I will bring to bear on the claims comes from several research projects, some investigating production of location expressions, others investigating comprehension of location expressions.

2. Constructing Location Expressions

2.1. Locating One of Two Identical Targets

You are Secret Agent U working with Secret Agent Z. You are in different places in the lobby of a hotel looking at a pair of potted palms, identical except that one hides a cache of stolen diamonds. You need to write a brief message on your secret communicator to let Z know which palm hides the diamonds. This was the 'tell other' task that Mainwaring, Ogishi, Schiano, and I presented to participants. (Mainwaring, Tversky, Ogishi, and Schiano, forthcoming).In the 'ask other' variant, U asked Z a yes/no question whose answer would reveal the target. In the 'tell self' variant, Agent U wrote a brief reminder for U to use later. Participants used depictions to understand the scenes. The depictions were maps that provided the elements that could be used to identify the location of the target. The depictions always included the relative locations of U and Z and of the identical objects. U and Z were either displaced 180 degrees facing each other or displaced 90 degrees. The pair of objects was either lined up with U or not. The depictions could also include a landmark that was informative, that is, closer to one object than the other, or a landmark that was uninformative, that is, equidistant between the objects. Finally, the depictions could include the direction of north, or both a landmark and an indication for cardinal directions. This task was expanded from a task used by Schober (1995); it was developed to investigate whether providers of information favored their own perspective or that of the receivers of the information. We also wanted to know if information providers would select a perspective that was neither their own nor their interlocutors when the presence of landmarks or cardinal directions enabled neutral perspectives.

As expected, participants used all the locating devices we expected them to use, and invented some we hadn't expected as well. One surprise was using Agent Z's perspective in the 'tell self' task. Since Agent Z wasn't part of the communication, it can't be an example of using the other's perspective. Rather, participants seem to be using Agent Z as a landmark, coding the target as the one that is closer to Agent Z. We then began examining the other tasks and layouts more closely and found that almost wherever possible, information providers chose to use *near* a landmark, whether the landmark was an object or an interlocutor in the scenario. When scenarios contained no landmarks or cardinal directions, all other things being equal, information providers tended to favor the perspectives of the information receivers in the tell other task. This is presumably because the information provider wants to ease the cognitive burden of the information receiver. The notable exceptions to using the other's perspective were when *near me* or *in front of me* could be used, especially to avoid the direction terms *right* and *left*.

Not only was using a landmark without a direction term preferred in most situations where it was sufficient, but also *near* was preferred to *far*. This is despite the fact that by considerations of markedness, *far* should be preferred, as the term for the dimension is 'distance'. Why was *near* preferred? We think it is because there is less ambiguity about where *near* is. *Near* is a region surrounding and close to the landmark. The region defined by *far* radiates in increasing and unbounded distance in all directions from the landmark.

Wherever reasonable, then, providers of information chose to express location-using closeness to a landmark rather than specifying a direction. When they did need to specify a direction, they preferred *in front* to *left* or *right*. *Front* and *back* are with reference to asymmetric, easily distinguished parts of the body as well as easily distinguished parts of objects (e.g. Franklin and Tversky, 1990). By contrast, the left side of a body or an object is usually quite similar to the right side. *Front* and *back* are easier to discern, hence easier labels to produce and to comprehend. In tasks requiring pointing to or remembering directions around the body or around a doll, participants used a larger region for front and back than for left and right (Franklin *et al*., 1995). Participants in the secret agent study also used cardinal directions; for those, there was no preference of one axis, north–south or east–west, over the other.

2.2. Descriptions of Environments

In a series of experiments, Taylor and Tversky (1992a; 1996) gave participants one of a variety of maps to memorize, including maps for a museum, a park, a theme park, a recreational area, a zoo, a convention center, a town. After studying the map, participants were asked to write a description of the environment the map depicted. One variable of interest was the perspective of the description. Participants used one of two perspectives, or, surprising the linguists, a combination of both. In a **survey perspective**, a fixed viewpoint above the environment was established and landmarks were described relative to each other in terms of north, south, east, and west. A survey perspective sets a scene and locates landmarks within it. In a **route perspective**, the viewpoint was the changing viewpoint of a traveler in the environment; landmarks were located with respect to the traveler in terms of left, right, front, and back. A route perspective provides procedures for traveling through the environment, locating landmarks en route. In both perspectives, the direction terms preferred are categorical, drawn from the sides of the body or the cardinal directions. Distances are rarely given; instead, action points on paths are delineated by landmarks or intersecting paths. Overall, environments that had a single path and landmarks at a single size scale attracted relatively more route perspectives and environments that had multiple paths and size scales attracted relatively more survey descriptions. Characteristics of the environments, then, determined description perspective, at least in part.

2.3. Depictions and Descriptions of Routes

Lee and I (Tversky and Lee, 1998, 1999) stopped students outside a dormitory and asked them if they knew where a popular fast-food joint was. If they did, they were asked to write down directions or else draw a map to the place. Both maps and directions could be segmented according to a system developed by Denis (1997) into sequences of four elements: start **point**—you exit the train station; **reorientation**—you turn right; **path/progression**—you go straight; and **end point**—until you reach the hotel. Moreover, the ways that the elements were represented corresponded so closely that automatic translation between route depictions and directions seems possible. This is because although the route maps could allow analog representation of landmarks, paths, turns, and distances, they tended to code these discretely. For example, intersections or turns were drawn as approximately 90 degrees; the deviations in drawing did not correspond to deviations of the intersections from 90 degrees in the world. Similarly, in language, turns were conveyed as *take a*, *make a*, or *turn* without specifying degrees of turn. Paths were drawn as straight or curved; the curve was again approximate, not accurate. Descriptions also made a two-way distinction. For straight paths, the corresponding description was *go down*; for curved paths, the corresponding description was *follow around*. Even distance was only schematically represented in the route maps; in fact, larger space seemed to be allocated to more complex intersections, not necessarily to longer distances. Longer distances were often indicated by broken lines, an explicit violation of the spatial correspondence of distance in representing space reflecting distance in real space. In both depictions and descriptions, distance was delineated indirectly and inexplicitly as the path between two readily recognized landmarks. That is, the information provider seems to be relying on the ability of the receiver to recognize and keep track of landmarks, not explicit distances.

For both the extended descriptions of environments and the route directions, people used landmarks and paths. The paths were typically defined by landmarks that demarcated a start point and then either a direction or another landmark serving as an end point. On occasion, number of blocks was used. Thus distance was rarely expressed directly, and even more rarely in conventional units.

Presumably, had degree of turn or curvature or distance been critical to wayfinding, further distinctions would have been made in both depictions and directions. Nevertheless, the hedges given in actual discourse are rarely in conventional units, but rather in descriptors like *a little bit to the left* or *a wide turn* (e.g. Franklin et al., 1995). The important point is that both depictions and directions for routes are schematic, they tend to be categorical rather analog, they discard information not necessary for keeping on track. Furthermore, both maps and directions schematize in similar ways. The critical elements are landmarks and paths, with direction and distance indicated only approximately.

2.4. Directions as Fields and Lines

The use of direction terms in conveying which object is the target and in giving route directions differ in subtle ways. In the case of distinguishing the target object, terms like *front* or *west* designate a two- or three-dimensional region projected from an intrinsic side of a person or object. The target is said to be somewhere in that region. What is interesting is that there are shared expectations about the likelihood of different parts of the region (e.g. Carlson-Radvansky and Irwin, 1993; Franklin et al., 1995; Hayward and Tarr, 1995; Logan and Sadler, 1996; Morrow and Clark, 1988; Regier, 1996). The presupposition is that the language-specifying region combined with perceptual information in the environment is sufficient to pick out the target. In the case of providing route instructions, direction terms like *right* and *east* are used as **direction lines**; they describe one-dimensional paths in the environment.

This review of research on production of location expressions in a variety of situations has illustrated that people prefer to locate objects as near a known landmark if possible. If further information is needed, people add specific information conveying the direction from a landmark; they prefer directions that are easier to compute, that is, they prefer to use *front* or *back* to *left* or *right*. Here, direction terms define a region projected from the intrinsic sides of a body or object. For complex location expressions that entail several segments, direction terms serve to define paths from one landmark to another.

Thus, landmarks are the first thing used, alone, if possible, or with spatial relation terms if needed. Spatial relations expressing proximity, such as *near*, are preferred to spatial relations expressing direction. Directions with natural asymmetries are preferred to those without. The directions most used are gross categorical ones. Distances are most often defined by initial and final landmarks rather than quantitative units. When units are used, experiential ones, such as blocks or time, are preferred to standard units, such as degrees or miles.

3. Comprehending Location Expressions

3.1. Accessing Regions Defined by Directions

You are at the local museum of Natural History, visiting the Space Exhibit, which occupies two floors of the building. A large section of the second-story floor is missing so that large objects can be displayed in an open area spanning the two floors. You are standing on the second floor on the walkway that surrounds this open area. As you look in front of you, you see a handsome portrait of John Glenn. The portrait is a bright watercolor painting, signed by the artist. Directly to your left a large, rocky meteorite sits on a heavy table. It is about the size of a small boulder, but it looks to you to be dense enough to weigh a ton. Looking behind you, you see a map of the solar system, including the orbit paths of all the planets. The map is large enough for you to read from where you stand. You now look directly upward, where a full-sized spacesuit hangs from the ceiling. It is shiny and white, and it looks as though it was never used. Peering downward toward the first floor, you see a communications satellite directly below where you are standing. It consists of a metal ball, two feet in diameter, with a metal dish attached to it.

Participants studied descriptions like these that located six objects to each side of an observer, addressed as *you*. Then the observer was reoriented to face another object in the environment, and probed for the objects currently at various directions from the body by the direction terms, *front, back head, feet, left,* and *right*. Although the descriptions did not favor one area of space over another, reaction times to the probes differed systematically. Specifically, when the observer was described as upright in the environment, responses to head and feet were fastest, followed by responses to front and back. Responses to left and right were slowest (Franklin and Tversky, 1990). Franklin and Tversky explained these findings by the **spatial framework theory**. According to the spatial framework theory, observers construct a spatial mental framework consisting of projections of the axes of the body, and associate objects to it. The axes vary in accessibility depending on asymmetries in the body axes and in the environment, as in the upright case; the **head/feet axis** of the body is aligned with the only asymmetric axis of the world, the axis formed by gravity. The head/feet axis, like the **front/back axis** has salient asymmetries. Because of the confluence of body and environmental axis asymmetries, head/feet is fastest for upright observers, followed by front/back, followed by right/left, as the **left/right axis** lacks salient asymmetries. The reclining observer turns from side to front to side to back, so no axis of the body is aligned with gravity. In the reclining case, relative access times rely solely on body asymmetries. The front/back axis importantly separates the world that can be seen and manipulated from the world that cannot be seen or manipulated, and indeed, in the reclining case, retrieval times are fastest to front/back, and next fastest to head/feet.

These results have been replicated in several variants of the original situation, varying viewpoints (surrounded by objects or looking onto them), numbers of observers, whether person or room is moving, and mode of input (language, diagram, model, actual environment) among other things (e.g. Bryant and Tversky, 1999; Bryant, Tversky, and Franklin, 1992; Bryant, Tversky, and Lanca, 2000; Franklin, Tversky, and Coon, 1992; Tversky, Kim, and Cohen, 1999). The same or systematically varying pattern of retrieval times to the regions around the body emerges. These biases in comprehension converge with the biases in production observed in the earlier studies. The same regions, defined by directions from the sides of the body that are easier to produce are also the regions that are easier to comprehend.

3.2. Comprehending Extended Descriptions

As noted earlier, environments are typically described using either a route perspective or a survey perspective or a combination of both. Do mental representations established by the two perspectives differ? The results of Taylor and Tversky (1992*b*) for limited environments acquired through descriptions suggests that perspective-free mental representations can be established. Participants studied equivalent route or survey descriptions of environments. They then verified verbatim and inference statements from both perspectives. The inference statements contained information that was available in the texts, but was not explicit, such as spatial relations from a different viewpoint. Verbatim statements in the perspective not read were equivalent to inference statements. On the whole, error rate was low. Nevertheless, participants were faster and more accurate on verbatim statements than inference ones. For the inference statements, there were no differences in either reaction time or errors for statements from the read perspective and the other perspective for both perspectives. This suggests two generalizations about representing discourse. First, there seems to be a representation of the text *per se*, conferring an advantage to verbatim statements. Second, there seems to be a spatial mental representation that is perspective-free. Such a representation may be like an architect's model, more general than either a route or a survey perspective, one that allows the taking of either of those two perspectives with equal facility.

In research in progress, Lee and Tversky are investigating the online construction of spatial mental models from route and survey descriptions. In a typical experiment, participants read several sentences describing an environment from one perspective and then a target sentence from the same or other perspective. This was followed by true/false statements in both perspectives. Switching perspectives did exact a cost in reading time and in errors. Thus, perspective matters in establishing mental representations, but may not matter once mental representations are well established, at least for these relatively small and well-learned environments. Follow-up studies using sentences that mix perspectives indicate that the largest cost in perspective switching is a change in direction terms, rather than a change in reference object.

Conclusion

The results from comprehension tasks converge with those from the production tasks: Landmarks are relatively easy and direction relatively difficult. *Left* and *right* are more difficult than *front* and *back* or *head* and *feet* or *above* and *below*. There is no strong evidence that egocentric reference terms are easier or harder than environmental reference terms, though some languages avoid egocentric reference terms (Levinson, 1996). Conventional direction and distance units are avoided. In fact, explicit distance is rarely provided. Rather, the beginning and end of a path is indicated by the landmarks located at the beginning and end.

What makes an expression 'easy'? Language habits and conventions develop in social situations where speakers and hearers collaborate in conveying information (e.g. H. Clark, 1996). Through these interactions, users have the opportunity to learn what kinds of expressions are effective and what kinds are not. As a result, common ground is established and language becomes more efficient. Many of the same phenomena occur when a person communicates with him or herself, that is, when a person commits something to memory and then retrieves it. Failures of communication to self or other warn what expressions should be avoided and successes of communication teach what expressions should be adopted. Efficient language capitalizes on expressions that are relatively easy and accurate for speakers to produce and for listeners to comprehend. For spatial language, ease and accuracy depend on human information-processing skills but also on spatial knowledge. The spatial knowledge that is more readily encoded and represented in everyday interactions in the world is undoubtedly knowledge that is more readily conveyed in language. Thus, linguistic habits and conventions can reveal spatial knowledge (e.g. Talmy, 2000). What the work on language suggests is that people are especially adept at recognizing and remembering landmarks and paths and that directions and distances are approximate, and wherever possible, delineated by landmarks and paths rather than quantitative units. These linguistic practices in fact reflect spatial knowledge (e.g. Tversky, 2000*a*, 2000*b*).

8

Ontological Problems for the Semantics of Spatial Expressions in Natural Language

PIERRE GAMBAROTTO and PHILIPPE MULLER

Abstract

This chapter presents a model of direction representation based on regions of space as primitives. The model is applied to the semantics of spatial expressions. It is shown that spatial relations expressed in language involve factors beyond geometry, that relate to the nature of the objects and the physical concepts involved.

Introduction

Reference to space in natural language is the focus of this chapter: what concepts are necessary to account for the semantics of spatial prepositional phrases or motion verbs, and what does this semantics reveal about the nature of cognitive space (the way spatial information is mentally represented)? This is not a recent issue as it is linked to a tradition of philosophical questioning about space in general, which can be dated back to the works of early Greek mathematicians. We will try to review here how this issue can be related to the issue of space in natural language, and we will focus on the question of orientation.

It is possible to isolate two problems in describing the meaning of spatial expressions. The first one is the problem of the concepts involved: is spatial meaning only the expression of geometrical knowledge, and if so, what kind of geometry is involved?

Geometrical concepts seem to be of prime importance when spatial knowledge is expressed (Miller and Johnson-Laird, 1976), and such concepts have even been considered sufficient to represent the semantics of certain natural language expressions (Leech, 1969). On the other hand, Talmy (1983), Vandeloise (1991), Herskovits (1986), Aurnague and Vieu (1993) have shown convincingly that other factors have to be taken into consideration to represent the meaning of spatial expressions and that they cannot easily be reduced to geometric concepts..

The second problem when dealing with spatial expressions is to explain how their meaning is organized—how can different kinds of spatial information be combined to produce well-formed natural language expressions, abiding by the compositional principles of semantics?

These two problems in turn raise the question of what objects are primitive in spatial cognition and serve as a medium for communication. By this we mean, what in our perception are the **ontological primitives**? To answer this question implies that we have isolated the important concepts that are part of the geometry of common sense, or in other words, the geometry of perception. Although the terminologies often differ, three components can be distinguished on the basis of studies of natural language expressions in various languages (Hays, 1989; Herskovits, 1986; Talmy, 1983). The most basic kind of knowledge is topological, expressed in prepositions such as *in, on, against*. Intuitively, **topology** only concerns the concepts of **inclusion** and **contact**. Then comes **orientation** (expressed by a prepositional phrase such as *to the left of*) and **distance** (*far, close, one meter from*).

From an ontological point of view, the issue is to choose the primitive objects that are used with such concepts. For instance, there has been some debate over the opposition between **axes** and **vectors** as the ontological basis of spatial cognition (Zwarts, 1997): axes would allow us to code relative orientations between spatial objects, whereas vectors would code both orientation and distance information. Both these options, however, take for granted a geometrical background inherited from **Euclidean geometry**. The primacy of such a framework from a cognitive point of view has been questioned in the literature. Dating back to Whitehead (Whitehead, 1929), a number of studies have proposed alternative frameworks to code spatial knowledge, claiming a closer approximation to human perception of space.[1] Much of the work in this area takes regions as primitive objects. It is now accepted that topologies can be reconstructed with reference only to regions of space corresponding to physical objects, and without resorting to such abstract entities as **points, lines**, or **vectors**. Although representing orientation or **distance** seems much less obvious from this perspective, promising attempts have been made towards a complete geometry based on regions (e.g. Aurnague and Vieu, 1993).

---

[1] In philosophy (Roeper, 1997; Varzi, 1996), mathematics (Gerla, 1994; Tarski, 1969), and more recently in Artificial Intelligence works striving to represent commonsense knowledge (Randell, Cui, and Cohn 1992; Borgo, Guarino, and Masolo, 1996).

In this chapter we make a case for a semantic representation of spatial expressions using a **region-based ontology**. We will show that a semantics based on regions is adequate to account for the meaning of prepositions. We will present a **multilevel semantics**, that is a semantics including geometric concepts and other concepts based on the nature and functions of objects referred to in natural language expressions about spatial situations. At the end of the chapter we will propose a list of problems for any semantic representation of space that, to date, seem to be unsolved.[2]

1. Reference to Space in Language

In this section we will review some of the concepts that seem necessary to account for reference to spatial situations in language, and the kinds of representational problems these concepts raise.

1.1. Spatial Relations

Space is referred to in many languages by using the following syntactic structures:[3]

- prepositions, selecting physical entities as NPs (*on the table, over my head,* ...);
- verbs: motion verbs (*run, cross, walk,* ...) or verbs describing static situations (*hang*);
- nouns (*top, corner,* ... as in *the top of the mountain, the corner of the table*);
- adjuncts of various kinds (*five miles*);
- adjectives describing spatial properties (*high*).

We will focus here on the class of expressions that are studied most: spatial prepositions. The conventional approach is to regard prepositional phrases as describing a relation between two or more objects, sometimes assuming a specific reference frame (Herskovits, 1986; Talmy, 1983).

---

[2] What is presented here is a synthesis of the results of an ongoing project on the semantics of spatial expressions launched by André and Mario Borillo (Aurnague *et al.*, 1991), followed by seminal works by Michel Aurnague and Laure Vieu, and continued in studies by, besides the authors of this chapter, Pierre Sablayrolles and Laure Sarda.

[3] We take examples from English for the sake of illustration; equivalent expressions covering the same concepts in other languages can be found in numerous studies; see Wierzbicka (1996); Talmy (1983).

The way a relation between two objects is specified in language is often underspecified with respect to the exact geometric situation involving the two objects, focusing on only some of its characteristics (Talmy, 1983).

(1) a. The fish is in a bowl.

Here the only information from a geometric point of view is of a topological nature: nothing is known about the precise location of the fish within the bowl, only that the region of space occupied by the fish is included in some interior part of the space determined by the bowl.[4]

It is important to note that such a description could have been further refined to any desired level of precision, as exemplified below, although this can be a disadvantage for communication, except, perhaps, among mathematicians:

(1) b. The fish is swimming in a bowl, near the bottom.
  c. The fish is swimming in a bowl, and its center of gravity has the trajectory (in cm) in a system of orthogonal axes:
  $x = 10.0 * \cos(t)$
  $y = 10.0 * \sin(t)$
  $z = 2$
  with an origin of the coordinate system at the center of the bottom of the bowl.

The constraints on the semantic acceptability of linguistic descriptions with respect to a particular situation do not only depend on geometric factors, as was convincingly shown by Vandeloise (1991). Other contributing factors to the semantics of prepositional phrases are: the nature of the objects involved (salient properties, privileged parts), their functions or canonical use, in addition to pragmatic factors depending on the context of the utterance. For instance, sentence (2) can describe various situations, depending on the frame of reference, because of a property of the object *car*:

(2) The driver is behind his car.

If the semantics of *behind* implies that the position of the two objects related (the car and the driver) is relative to a direction, only the context of utterance can determine that axis, because there are two possibilities: either (a) the car is between the driver and the speaker of (2) (the axis is given by the point of view of the speaker relative to the car), or (b) the driver is at the rear of the car

---

[4] This space is of course different from the space occupied by the physical object, and is related to the **interior** of the object. This is not just a geometrical notion, as shown in Vandeloise (1991).

(the axis is given by the intrinsic orientation of the car, which has a front and a rear).

The nature of the objects related can also change the meaning of an expression. Compare for instance, (1) and (3). In (3) the flower is not necessarily *entirely* included in the space determined by the bowl. This means that there are constraints on the two objects related by the preposition, and not only on the NP in the PP, as is often assumed by approaches where the meaning of a spatial PP is assumed to refer to a region of space (Leech, 1969).

(3) The rose is in a bowl.

Finally, certain concepts that are part of our commonsense knowledge (what has sometimes been called **naive physics** (P. Hayes, 1985)), such as containment and support, are necessary to explain such differences as between (4a) and (4b), where the same preposition is used in different geometrical configurations:

(4) a. Le verre est sur la table. ('The glass is on the table.')
    b. L'affiche est sur le mur. ('The poster is on the wall.')

Here, it's the notion of physical support that links the two uses of the preposition *sur* (*on*): in both cases, the first object is physically related to the second in a way that opposes gravity.

The same kind of difference can be observed in relation to motion verbs:

(5) a. La voiture a traversé le mur. ('The car went through the wall.')
    b. La fourmi a traversé le mur. ('The ant crossed the wall/went through the wall.')

Here the scale of the figure (ant or car in motion) determines the context and hence the spatial interpretation of the expression *traverser le mur* (and (5b) is thus ambiguous).

1.2. Representational and Methodological Problems

The facts reviewed in the previous section raise a number of issues concerning the representation of space, and question the adequacy of classical spatial representations (e.g. as used in physics or in Artificial Intelligence's early attempts to capture commonsense related notions, such as Forbus (1983)). **Classical spatial representations** assume that space is a **Euclidean affine space** isomorphic to $\mathbf{R}^3$, the set of all ordered triples of real numbers. In such a view, the semantics of, for example, prepositional phrases would be sets of points in $\mathbf{R}^3$ with specific constraints on the nature of those sets. In the same way, motion is considered as a function from time (R) to space (R^3). It is therefore based on a direct transposition of physics concepts. Moreover, studies on the representation of prepositions often assume that a prepositional phrase determines a region of space (a location) (see Zwarts, 1997, for a review).

The classical notion of spatial representation raises two problems. The first is an adequacy problem: how easy is it to relate a point-set model to the actual use of language, bearing in mind what was observed in section 1.1, namely, that the meaning of a preposition depends on the objects related more than on mere geometrical information? The second problem was raised by Zwarts (1997), and concerns the lack of **compositionality** of a **region-only semantics** based on sets of points for preposition and spatial adjuncts. We have claimed that the meaning of a preposition is more likely to be a relation between *two* objects (and the regions of space they determine), so these two problems boil down to a single one: the ontology we choose.

The ontological debate also raises a methodological problem: how are we to prove that such and such a representational framework is closer to the 'reality' of our cognition? We believe two kinds of answers can be given. The first answer is tentatively given by psycho-linguistic experiments, such as those described in Taylor and Tversky (1992*b*). Another, less classical means, is to check the inferences that are made possible by different models and to compare them to what intuition would tell us. This approach has been followed by numerous studies in our research group, a prototypical example being the study of French orientation prepositions by Aurnague (1995). Eventually, the two methodologies can meet, through experiments testing the hypotheses developed in a framework where inferences are checked against intuitions.

We will now present ways of representing spatial information in a framework departing from Euclidean geometry, and see how it can be used, along with functional information, to model the semantics of spatial expressions.

2. Region-Based Ontologies

As shown above, spatial structures handled in language are mostly of a relational nature. Studies in Artificial Intelligence at first focused on classical structures of space derived from mathematics (absolute spaces based on **axes** and **coordinates**) in domains such as robotics or vision. More recently, research has taken an interest in **relational spaces** for knowledge representation. The reason for this is that there are two main drawbacks in assuming absolute spaces. First, the position of each entity in a coordinate system has to be represented precisely. Second, a coordinate system presupposes the choice of a length unit, making it hard therefore for representations in such a space to adapt to different degrees of granularity, an essential part of reference to space in language. Positions of objects are usually expressed in a rather imprecise and incomplete manner. Relational spatial structures make it theoretically possible to represent position while preserving their intrinsic underspecified nature. Although it is possible to leave a coordinate-based representation underspecified (using inequalities for instance) it goes against the whole purpose of the **absolute space approach**, which is to derive a precise calculus on coordinates.

Among the numerous publications that have chosen the relational approach, it is possible to differentiate a number of trends, based on the choice of the ontological primitives used. For instance, Frank (1992) and Freksa (1992) consider points as primitive entities. **Objects** are therefore represented as punctual, and distance and orientation are considered only in cases where points can represent objects. A second approach is based on the work of Allen (1984). Allen's work on time intervals can be generalized to greater dimensions. Objects can thus be defined as sets of intervals, depending on the dimensions of the space considered. The obvious shortcoming of this method is to limit the possible shape of objects to parallelepipeds, which is clearly not realistic. We take a third approach here, as it seems to us that it overcomes some of the problems of the previous approaches in natural language semantics. Concrete objects are associated with extended bodies, called **individuals** or **regions**. Basic **topological relations** are then easy to axiomatize, following the work of Lesniewski (1992) or Clarke (1985). We will then see how to represent directions, orientation, and distance between objects based on such axioms.

2.1. Topology

It is possible to consider two types of relations. **Mereological relations** refer to part-whole relations, derived from inclusion. These relations can easily be axiomatized with a single primitive P, interpreted as follows: Pxy means x is a **part of** y:[5]

A1) $\forall x\, Pxx$ (everything is a part of itself)
A2) $\forall x,y\, (Pxy \wedge Pyz) \rightarrow Pxz$ (a part of a part of a whole is also a part of the whole)
A3) $\forall x,y\, Pxy \wedge Pyx \rightarrow x = y$ (if two entities are part of each other, then they are equal)

---

[5] We have adopted a notation without parenthesis for predicates, and we omit universal quantifiers taking scope over formulas, for the sake of conciseness. We use here a classical first-order language with equality.

Other mereological predicates can then be defined:

D1) $PPxy \equiv_{def} Pxy \wedge \neg Pyx$ (proper part)
D2) $Oxy \equiv_{def} \exists z\ Pzx \wedge Pzy$ (overlap: x and y have a part in common)
D3) $POxy \equiv_{def} Oxy \wedge \neg Pxy \wedge \neg Pyx$ (proper overlap)

Topological relations are more general than mereological ones. The primitive chosen for the axiomatization is generally C. Cxy is interpreted as 'x and y are **connected**'.

A4) $\forall x\ Cxx$ (everything is connected to itself)
A5) $\forall x,y\ Cxy \rightarrow Cyx$ (connection is symetric)
A6) $\forall x,y\ (\forall z\ (Czx \leftrightarrow Czy)) \rightarrow x = y$ (if x and y are connected to exactly the same things then they are equal)

Mereology is subsumed by topology: P can be defined using C only.

D4) $Pxy \equiv_{def} \forall z\ (Czx \rightarrow Czy)$ (x is part of y whenever everything connected to x is also connected to y)

Thus, the expressivity of topology is superior: the relation of **contact** (**tangential part** and **external contact**) can only be defined in topology.

D5) $ECxy \equiv_{def} Cxy \wedge \neg Oxy$ (external connection: x and y 'touch' each other)
D6) $TPxy \equiv_{def} Pxy \wedge \exists z\ (ECzx \wedge ECzy)$ (tangential part, 'touching' the border of y)
D7) $NTPxy \equiv_{def} Pxy \wedge \neg TPxy$ (non-tangential part, or **interior part**)

All these relations are illustrated in Fig. 8.1.

The notion of **interior** can then be introduced:

A7) $\exists y \forall u (Cuy \leftrightarrow \exists v (NTPvx \wedge Cvu))$ (y, noted ix represents the interior of x)

Other topological concepts can now be described, such as **open/close objects**, and also **connectedness**, or even a notion of **weak contact** (Asher and Vieu, 1995), which models the concept of physical contact.

These notions are necessary (if not sufficient) to describe the semantics of prepositions that involve physical inclusion or contact (*in, on, against,* ...). For instance to express the spatial location of the fish in the bowl of example (1a), we could say that: P(fish,interior(bowl)), where the function interior()

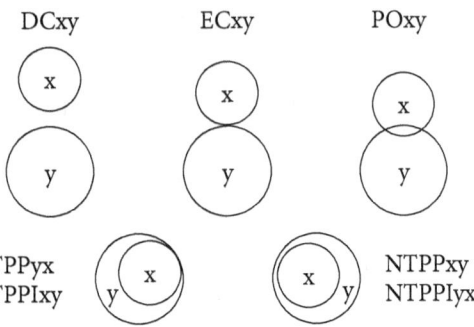

FIG. 8.1 Mereo-topological relations.

denotes the containing part of an object (and is different from the notion of **topological interior**, see Herskovits (1986), Vandeloise (1991), or Aurnague and Vieu (1993)).

We have not detailed all the properties of the relations we have introduced, as we only want to insist on the vocabulary they provide, and their approximate designation. A good discussion of the precise topological properties one would want for spatial representation can be found in Varzi (1996).

2.2. Orientation and Distance

The representation of **orientation** requires a complex mechanism. As pointed out, for instance, by the work of Aurnague (1995), the problem cannot be solved at the geometrical level alone. The core idea behind orientation is the notion of direction. More precisely, orienting an object is equivalent to choosing and giving a particular name to a specific direction of the object. A **direction** is typically defined by a pair of points O and E (**Origin** and **Extremity**). The opposite direction is obtained by exchanging O and E. In order to build orientation on a region-based geometry, it is necessary to redefine the notions of point, origin, and extremity. Two strategies can be followed. In the first one, the introduction of points in a **mereo-topology** is done by considering a point as a set of regions, i.e. the exact opposite of the classical view, as for example used in Schmidtke, Tschander, Eschenbach, and Habel (Ch. 9). Points are built from already existing objects, and thus do not require extra vocabulary. The second strategy is to introduce new regions, representing object-parts (extremities). We will see how both these strategies can be developed.

2.2.1. *Building 'Points' from Regions*

If we consider the intuitive use of the notion of '**point**' in natural language, as in *the point of the pen*, we notice that it never denotes a dimensionless entity, but rather a part of an object that seems minimal in a given context, and more importantly, whose internal structure is irrelevant in that context. The same kind of property seems to hold for objects involved in specifying direction. For example, when one says *look in the direction of the open window in that building*, we do not consider every part of the window. The window itself is seen as punctual; because we are not interested in anything other than looking at the object as a whole.

In our view, a useful notion of a point is that of an equivalence class of regions that have something in common. Let us consider a set of atomic regions (regions without parts). These regions can be considered as 'points' in the sense we have just described. If we consider points to be the minimal parts of all entities there is a problem. What if there are no minimal parts, because entities can be indefinitely subdivided into smaller parts? In this case we say that a point is the (possibly infinite) set of all regions that have a part in common. The finite case is graphically illustrated in Fig. 8.2. A formal definition of a point from a set of regions is what is mathematically known as an **ultra-filter** on the set of regions:

let 'a' be a set of regions such that:
- $\forall\ x,y \in a$ Oxy (x and y overlap) and the intersection x.y of x and y is also in a
- $\forall\ x \in a, \forall y\ Pxy \rightarrow y \in a$ (every part of a region of a is also a region of a)
- a is not empty
- a is maximal for the previous properties

then 'a' is a point.

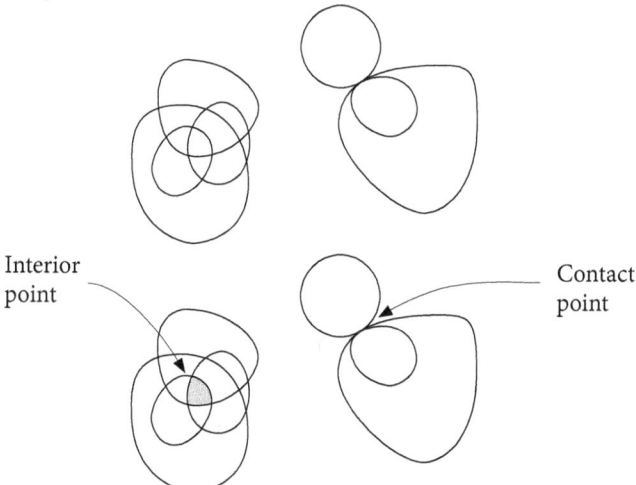

FIG. 8.2 Points constructed from regions, and an example of an induced direction (right).

Given a finite world, a point amounts to the set of all regions containing a given atomic region. Otherwise it is the mathematical limit of a set of embedded regions.

There is something else we want to consider as a point in a topological theory: the place where two entities 'meet' or 'touch'. We can speak, for example, of *the point where the sun meets the earth* and try to *go in that direction*. Theoretically, we can formalize such points in a similar way: whenever two regions are in external contact, they define a point. Pairs of touching regions nested in another pair of touching regions must necessarily denote the same 'touching' location to be part of the same point (see Fig. 8.2). Thus a formal definition as given in Aurnague and Vieu (1993): an external point a is a set of regions such that:

- $\exists x, y \in a$ Ecxy (at least two regions of a are 'touching')
- $\forall x, y \in a$ (Oxy $\wedge$ x.y$\in$a) $\vee$ $\exists z, t \in a$ (Pzx $\wedge$ Pty $\wedge$ ECzt) (if two regions of a intersect, either their intersection is part of a, or a contains two subparts z of x and t of z which touch each other)
- $\forall\ x \in a, \forall y$ Pxy $\rightarrow$ y$\in$a (any y of which a region x of a is a part is also a region of a)
- a is maximal for the previous properties.

It is not important to understand the technical details of this rather complicated definition, but note that Asher and Vieu (1995) use it to show the correspondance between topologies based on regions and on points. It is enough for our argumentation to see these points as equivalence classes of touching regions.

We have thus defined two kinds of points from a mereo-topological theory of regions of space. It is now quite straightforward to define direction in the usual way (a **direction** is a pair of points, an **oriented direction** is an ordered pair of points). Predicates of distance are also present in the work of Aurnague and Vieu (1993): K and T are ternary predicates, meaning respectively, if a, b, and c are arguments (points as defined above): K(a,b,c) means 'a is closer to b than c' and T(a,b,c) means 'a is between b and c'.

What is important from an ontological point of view is the construction, from a set of perceptually relevant regions of space, of derived notions that are commonly seen as separated or abstract. We have presented an approach in which points are *secondary* entities. This, however comes with a price: points are second-order items and thus reasoning on the basis of such a logical theory cannot be automated as easily as in a theory of regions alone.

2.2.2. *Directions as Special Parts of Objects*

If we want to encode object orientation, it is possible to isolate special parts ('extremities') and to define directions as pairs of extremities of an object. A first solution in defining extremities is to use the notion of **limit**. Considering a region as the spatial referent of a concrete object (and closed in the topological sense), a limit is defined as a tangential part having an empty interior, i.e. containing no closed region as a part:

> Empty(x) $\equiv_{def}$ $\forall$yPxy→OPy (x is empty means either it has no parts or they are open (OP))
> Lim1(x,y) $\equiv_{def}$ Empty(ix)∧TPxy∧$\forall$z (TPzx→TPzy) (x is a limit of y if its interior is empty, it is a tangential part of y and any tangential part of x is also a tangential part of y).

To locate regions with regard to one another on the same direction, an extension of Allen's (1987) 13 relations is introduced: R(x,y,D) describes the configuration between the projections of the regions x and y in the direction D (these projections are therefore similar to intervals).[6]

> Ext(y,x,D) $\equiv_{def}$ Lim1(y,x)∧$\forall$v((Pvx∧¬Pvy) → <m(v,y,D))

Here y is an extremity of x in the direction D if y is a limit of s and every region included in x and not in y is before (<) or meets (m) y in the direction D (the notation <m indicates that one of < or m holds between v,y, and D).

Exts(y,z,x,D) completes Ext : y and z are extremities of x, respectively for the direction D and −D. This framework provides the necessary tool to handle basic orientation.

Figure 8.3 illustrates what it means for a region to have a privileged part as an extremity.

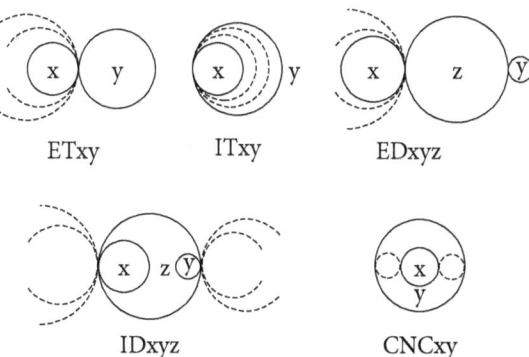

FIG. 8.3 Relations between spheres.

---

[6] Allen's classical relations are = , < (before), m (meets), o (overlap), s (starts), d (during), f (finishes), and their converse relations.

The difference between this representation of orientation and the preceding reconstruction of points is that the latter will specify only some of the properties of orientation (as directions are considered as primitives and are related only to some parts of objects, namely extremities), but in a first-order theory suitable for inference-testing in a formal way, see section 3.3 (which could be automated). From the perspective of natural language representation, the semantics of lexical items is generally expressed in terms of first-order properties.

The solution above presupposes the existence of particular regions in order to define extremities. Another solution developed in (Dugat, Gambarotto, and Larvor, 1999), is to introduce new specific regions. These regions are then used to define geometric concepts such as direction and angle. In order to have a rich vocabulary, the regions introduced are spheres. The main advantage here is the possibility of drawing on previous work. Tarski (1969) has proposed a mereology based entirely on spheres. Due to their isotropic nature, spheres are easily handled shapes, and Tarski gave several definitions involving two or three spheres. Each definition characterizes particular positions, and these are depicted in Fig. 8.3.

$ETxy \equiv_{def} \neg Oxy \wedge ((\neg Ouy \wedge \neg Ovy \wedge Pxy \wedge Pxv) \rightarrow (Puv \vee Pvu))$ (x is externally tangent to y)

$IT \equiv_{def} PPxy \wedge ((Puy \wedge Pvy \wedge Pxu \wedge Pxv) \rightarrow (Puv \vee Pvu))$ (x is internally tangent to y)

$ED \equiv_{def} ETxz \wedge ETyz \wedge (\neg Ouz \wedge \neg Ovz \wedge Pxu \wedge Pyv \rightarrow \neg Ouv)$ (x and y are externally diametrical wrt z)

$IDxyz \equiv_{def} ITxz \wedge ITyz \wedge (\neg Ouz \wedge \neg Ovz \wedge ETxu \wedge ETyv \rightarrow \neg Ouv)$ (x and y are internally diametrical wrt z)

$CNCxy \equiv_{def} EQxy \vee (PPxy \wedge (EDuvx \wedge ITuy \wedge ITvy \rightarrow IDuvy)) \vee PPyx \wedge (EDuvy \wedge ITux \wedge ITvx \rightarrow IDuvx))$ (x and y are concentric)

The following three definitions have been proposed by Borgo and Masolo (1996), and are useful to define **alignment of spheres**:

$BTWxyz \equiv_{def} \exists x' \exists y' \exists z' (CNCxx' \wedge CNCyy' \wedge CNCzz \wedge \neg' EDy'z'x')$ (x is between y and z)

$LINxyz \equiv_{def} BTWxyz \vee BTWxzy \vee BTWyxz \vee CNCxy \vee CNCyz \vee CNCxz$ (x, y, and z are aligned)

$SSDxyz \equiv_{def} BTWxyz \vee BTWyxz \vee CNCxy$ (x and y are on the same side of z)

Borgo and Masolo (1996) define a sphere using the primitive of **congruence** CG (CG xy means x and y are congruent). Given the definition of congruence and the sphere, it is rather simple to build a strict order over any set of spheres: if a sphere x is congruent to a sphere z and z is a part of the sphere y, then the sphere x has a smaller size than y. This order can be used to define metrical concepts, and by taking as primitive S (S x means 'x is a sphere') congruence can be defined.

The next step is to rebuild the geometrical notions of **distance** and **angle**. This is rather simple: a general construction in terms of a sphere is given by extending the classical definition on points to spherical regions. For instance, the distance between two spheres is given by the sphere aligned with and between the two spheres.

The difficult part is to be able to associate particular spheres to an object, in order to define specific directions. Dugat, Gambarotto, and Larvor (1999) have proposed an algorithm to associate a structured set of spheres to an object. Each sphere is in the object and maximal with regard to size, the convex envelope of the set roughly approximating the shape of the object. It is then possible to apply the notions of angle and distance already defined to extract notions similar to the extremities defined by Aurnague and Vieu. The main drawback of this method is the limitation to 2-D spaces. What this work shows is that it is possible to reconstruct a commonsense geometry from objects referred to in natural languages, and that this geometry will have the desirable properties we highlighted in the introductory sections: it deals primarily with objects and relations on them, possibly underspecified, and allows for representations at different granularity levels.

2.3. The Case of Motion

Some components of natural language spatial expressions are rather difficult to model, because they refer not only to the spatial dimension, but to the temporal one also. We gather such expressions under the **'motion expression'** label.

Motion is a very important concept in several fields ranging from philosophy to artificial intelligence, and the nature of motion and moving entities has long been debated. A very strong influence on this debate has of course been the **Newtonian framework** applied in physics, which supplements classical geometry by adding a separate temporal dimension. This construction entails a number of drawbacks: assuming motion to be obtained by a space-time combination involves combining difficulties from both concepts. Among the problems that have arisen by such an ontology of motion, we want to focus on the difficulty of deciding whether at a given instance an object is moving or not (Zeno's paradox), and on the major problem of identifying a given spatial object through time.

The classical ontological choice, in most axiomatic theories of objects and their parts, is to make a distinction between **continuants** and **occurrences**. The former model objects that persist through time ('normal' objects), whereas the latter stands for events or states (bounded in time). This duality entails difficulties when considering change in motion, especially in defining criteria for identifying objects whose parts or properties are changing through time.

This problem of continuity of continuants has been indirectly addressed in several works in the literature. Among them we can distinguish the solution of Galton (1997) that refines the notion of **temporal structure**.

A more radical treatment of the problem is to adopt an entirely new ontology where primitive entities are spatio-temporal. These entities can then be interpreted as the trajectories of physical objects and events. A number of authors have already suggested this approach. Among them P. Hayes (1985) clearly emphasized the expressive power of such a homogenous theory that considers events and objects as equivalent, thus directly solvting the problem of identity of object. Following this idea, the work presented in Muller (1998) adapts a classical mereo-topology to spatio-temporal entities. A classical mereological relation is thus interpreted in terms of space and time..

The relation of overlap is represented in a space-time of two dimensions (one for space, one for time) in Fig. 8.4. The spatial dimension of entities x or y are part of the horizontal axis and their respective temporal evolution is read from the vertical axis.

These spatio-temporal relations are supplemented with pure temporal ones. The chosen primitives are **temporal precedence** and **temporal connection**. From these two primitives the relations of temporal overlap and temporal inclusion can be defined, with an axiomatization inspired from the axioms of mereo-topology based on the primitive of **spatial connection** C (Aurnague and Vieu, 1993). The relations are represented on Fig. 8.5.

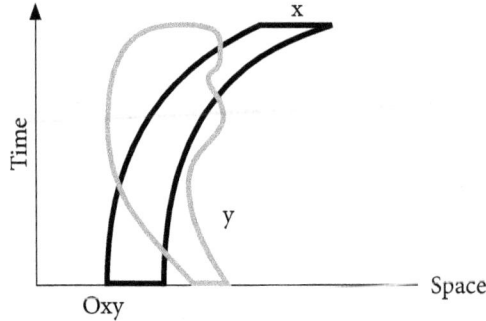

FIG. 8.4 A spatio-temporal interpretation of overlap.

Ontological Problems 159

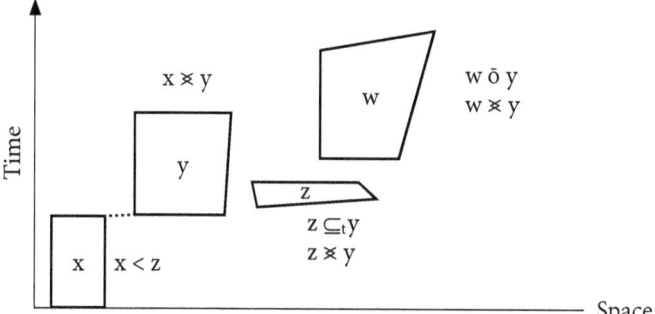

FIG. 8.5 Illustration of temporal relations.

Considering the parallel evolution of two spatio-temporal individuals that are continuous during the same time interval allows us to distinguish between a number of classes, represented in Fig. 8.6. Six classes of motion are defined: LEAVE, HIT, REACH, EXTERNAL, INTERNAL, and CROSS. An instantiation of each of these classes can be given by the following motion verbs : *leave, hit, reach,* and *cross* correspond to the topological aspect of the matching verbs, while INTERNAL can be associated with verbs such as *to drive* (around the country) and EXTERNAL with verbs such as *to avoid*.

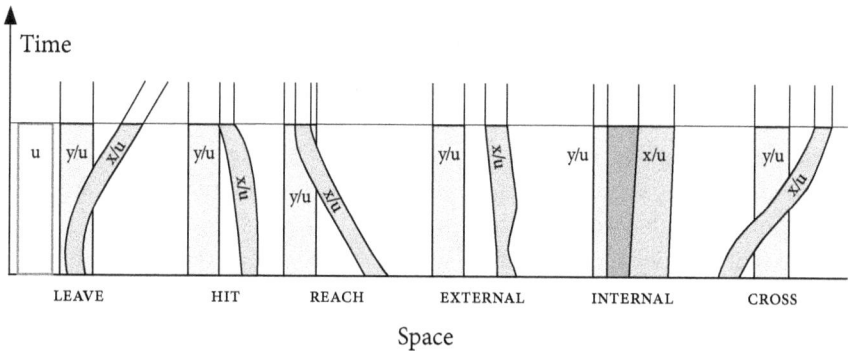

FIG. 8.6 Motion verbs.

3. Representing the Semantics of Prepositions

This section describes how the geometrical concepts introduced in the previous section can be used to represent the semantics of spatial prepositions together with **functional information**. As an example we will consider the

French prepositions *devant* and *derrière* (close in meaning to 'in front of' and 'behind'), and show which inferences seem to validate the adequacy of such a representation (Aurnague, 1995).

3.1. Functional Information

Encoding orientation in language is based on access to a **frame of reference** and involves specifying direction with respect to a frame of reference. A frame of reference can be determined by several factors, accounting for the semantic differences in the use of prepositions such as *devant/derrière*, which requires either an *intrinsic* frame of reference or a *deictic* one:

(6) Calvin est devant la télé. ('Calvin is in front of the TV.')
(7) Calvin est devant l'arbre. ('Calvin is in front of the tree.')
(8) La boite rouge est devant la boite bleue. ('The red box is in front of the blue box.')

Without prior context, the first of these examples is acceptable only if both the objects encoded by the NPs are facing each other, meaning they both have a privileged direction that indicates their respective front (**intrinsic orientation**), in this case, both objects have orientation determined by an **intrinsic frame of reference** (depending only on the objects involved). Example (7) is acceptable in two cases: when Calvin is facing the tree, which does not have any privileged orientation in French (intrinsic case), and when Calvin is between the speaker and the tree (**deictic orientation**). The exact frame of reference cannot be determined without knowledge of the context, and the sentence is ambiguous with respect to the spatial situation. Example (8) is only acceptable from the point a view of the speaker (**deictic frame of reference**), since boxes don't have orientations.

The first case where only intrinsic orientation is involved can be further refined according to the nature of the object. Compare, for instance, examples (9) and (6):

(9) Calvin est devant la voiture. ('Calvin is in front of the car.')

Example (9) is acceptable even if Calvin is not facing the car, since the situation does not involve a normal use of the other object (as in 6), which would constrain his position further.

What these examples show is the intricate dependence between the semantics of spatial prepositions and the properties of the objects related with such prepositions. It shows, moreover, the need for information other than geometry to deal with the semantics.

We will see below how to express these notions formally, within the framework provided in section 2. First of all, it is necessary to have a **two-level ontology**: one level for the objects referred to in natural language expressions, and one for their spatial referents. Whereas the previously introduced formalisms dealt only with the latter, we will now focus on the former. We will use the function 'sref' to denote the spatial referent of an object and express the link between the two levels. These two levels are necessary to distinguish an object and the matter it is made of, or when two entities are introduced into a discourse that occupy the same region in space (this difference is similar to the difference between events happening in time and the intervals of time of their occurrence).

3.2. Formalizing Functional Information

In order to characterize particular orientations in objects, a few definitions have to be introduced.

A **functional direction** can be defined as follows, relating it to a geometrical extremity as defined in section 2.

dir-ext(y,z,x) = D ↔ (part(y,x) ∧ part(z,x) ∧ exts(sref(y),sref(z),sref(x),D))
(D is the direction determined by the functional parts y and z of an entity x)

To define an intrinsic direction, we need only a few additional facts depending on the nature of the previously introduced direction D. For instance, an entity has an intrinsic upper-vertical orientation when it has a functional direction D and when this direction corresponds to the direction of gravity, pointing upwards:

orient-up(D,x) ≡$_{def}$ ∃y,z (dir-ext(y,z,x) = D ∧ can-use(x) ∧ (in-use(x) > dir-ext(y,z,x) = grav-up))

Can-use(x) means that x has a canonical use; the definition also says that when used canonically (in-use(x)), one can infer that D corresponds to grav-up. Here grav-up stands for the constant direction corresponding to gravity, pointing upwards. Likewise, a down-orientation could be defined with respect to gravity.

Frontal orientations involve more complex factors in order to account for cases shown in the previous section. As hinted at in that section, there are three prototypical cases of frontal orientations in French (cf. Vandeloise, 1991). The first case is where the frontal orientation is induced by what Vandeloise calls general orientation, for example the front in human beings, animals, vehicles, etc. This can be expressed as follows:

orient-front1(D,x) $\equiv_{def}$ $\exists$y,z (dir-ext(y,z,x) = D $\wedge$ orient-gen(x,D))

The orient-gen predicate could be defined according to specific properties of entities having such an orientation.

The second type of frontal orientation concerns those entities that have a **canonical use**. When used canonically the frontal orientation of the object is aligned with the user's frontal orientation (a car or a chair are typical examples).

orient-front2(D,x) $\equiv_{def}$ $\exists$y,z (dir-ext(y,z,x) = D $\wedge$ can-use(x) $\wedge$ $\forall$u,D'
(uses(u,x) $\wedge$ orient-front1(u,D')) $\rightarrow$ D' = D)

The third case of frontal orientation concerns those entities whose frontal orientation is opposed to that of their user when canonically used:

orient-front3(D,x) $\equiv_{def}$ $\exists$y,z (dir-ext(y,z,x) = D $\wedge$ can-use(x) $\wedge$
$\forall$u,D'(uses(u,x) $\wedge$ orient-front1(x,D')) $\rightarrow$ D' = $-$D)

We can now refer to an entity having a frontal orientation as falling into one of the three kinds of orientations:

orient-front(D,x) $\equiv_{def}$ orient-front1(D,x) $\vee$ orient-front2(D,x) $\vee$ orient-front3(D,x)

We can also define a symmetrical notion of 'back' orientation:

orient-back(D,x) $\equiv$ orient-front($-$D,x)

The last thing we need is a predicate In-sp(x,y,D) saying that an entity x is located in the **half-space** determined by another entity y and a direction D, that will give us a core geometrical notion of being *in front of* something. This can be defined in several ways using the concepts in section 2. A proposal made by Aurnague and Vieu (1993) makes use of the projection relations already introduced in section 2.2, which seem to have the desirable properties with respect to what can be inferred in context (see the following section).

In-space(x,y,D) $\equiv_{def}$ m_i > (sref(x),sref(y),D) (the referent of x is met (m_i is the converse of m or 'meet') or is after the referent of y in the direction D)

We now can define the two meanings of the preposition *devant*, as presented in section 3.1, i.e. intrinsic *devant* and deictic *devant*:

devant-i(x,y,D) $\equiv_{def}$ orient-front(D,y) $\wedge$ In-space (x,y,D) (x is intrinsically *devant* y with respect to direction D when y has D as a frontal orientation and x is in the space determined by y and D) devant-d (x,y,D) ≡$_{def}$ ∃s (orient-front (−D,s) ∧ s≠x ∧ s≠y ∧ speaker(s) ∧ devant-i(y,s,−D) ∧ In-space (x,y,D) (x is deictically *devant* y with respect to direction D when there is a speaker s, s has a frontal orientation (−D) and y is intrinsically *devant* s with respect to −D, and x is in the space determined by y and D)

The last definition says that in the deictic case, the underlying direction D is induced by the speaker.

This formal representation of the meanings of orientation prepositions in French was essential to model the non-geometric factors that are part of the semantics of these prepositions. It shows a way of dealing with information that constrains the objects that can be selected by a preposition in the case of static descriptions (e.g. in this case, the fact that some entities have intrinsic orientations). Using prepositions in combination with motion verbs would lead to different meanings, some of which have been investigated in Asher and Sablayrolles (1995).

3.3. Meaning as Inference

We will now consider the inferences that are supported by the semantics given, in order to check whether they satisfy our intuitions about the use of certain expressions. As we have already argued above, we take this as a necessary step in putting the cognitive adequacy of a semantic representation to the test.

We will consider the combination of the information contents of two sentences in which the prepositions represented above are used. These sentences are considered as partial descriptions of the same situation. We then compare in each case what additional information can be inferred from the representation to see if it matches our intuitions.

The first case we test is a combination of two intrinsic uses of the same preposition, as exemplified by (10), from Aurnague (1995).

(10) a. Le tabouret est devant le fauteuil. ('The stool is in front of the armchair.')
b. Le fauteuil est devant Max. ('The armchair is in front of Max.')

If the constants a, s, and m denote *the armchair, the stool,* and *Max* respectively, the meaning of the previous sentences amounts to:

devant-i(s,a,d$_1$) ∧ devant-i(a,m,d$_2$).

From the definition of these predicates, we have:

m$_i$ > (sref(s),sref(a),d$_1$) ∧ m$_i$ > (sref(a),sref(m),d$_2$)

Now, if we have no additional information available, we cannot infer anything. But if prior context somehow provides the information that the direction of the speaker coincides with that of the armchair (that is, if $d_1 = d_2$), then the axiomatic of Allen's relations implies that:

>(sref(s),sref(m),d_2) (the referent of s is after the referent of m in direction d_2).

Drawing on the fact that 'orient-front(d_2,m)', from devant-i(a,m,d_2), we can derive:

devant-i (s,m,d_2)

This corresponds to *le tabouret est devant Max* ('the stool is in front of Max'), from an intrinsic perspective).

Consider the following examples:

(11) a. Le tabouret est devant la plante.   ('The stool is in front of the plant.')
      b. La plante est devant le lampadaire.   ('The plant is in front of the light.')

The fact that neither of the entities involved has intrinsic frontal orientation (something that is coded at the lexical level) implies that *devant* is used deictically in each case. This could be reinforced with the expression *vu d'ici* ('from my point of view'). This means that the directions involved are in fact the same (i.e. the orientation of the speaker). So:

devant-d(s,p,d) ∧ devant-d(p,l,d)

In analogy to what we had before, we now have:

devant-d(s,l,d)

meaning that the stool is in front of the light from the vantage point of the speaker. This time no prior contextual information is needed.

Similarly, Aurnague (1995) shows that inferences from intrinsic/deictic combinations are possible. These cases all boil down to the transitivity of the relation devant(_,_,d), either intrinsic or deictic. He specifies conditions under which such inferences can be made (transitivity being invalid in the general case).

Only part of the actual semantics of *devant* has been represented here, however, since the utterances of expressions using this preposition also imply that the objects are relatively close to one another. Because the **granularity level** that determines what is 'close' is very context-dependent, it is not considered here. We have left aside such pragmatic considerations as they are not the focus of this chapter, but studies of pragmatic principles within the framework presented here can be found in Aurnague (1995),Vieu (1991), Aurnague and Vieu (1993). What we have presented here can be seen as a semantic core.

On a different level, it must be noted that the inferences we presented here could easily be validated experimentally, by asking people to solve the syllogisms presented above.

Conclusion

We have considered the problem of reference to space in natural language from a representational point of view. Starting from the study of spatial prepositions, it has been shown that spatial relations involve factors beyond geometry, concerning the nature of the objects and the physical concepts involved. By separating objects from the physical region they occupy in space, both factors can be modeled in a formal way, to be incorporated in a precise semantics of linguistic expressions. When considering the nature of spatial objects, we also had to reconsider the standard, point-based ontology of space (coming from classical geometry); in contrast to this approach we have built a model based on regions of space as primitives (following work in philosophy and mathematics). It is our feeling that this gives a cognitively adequate basis for representation, and one that has rarely been applied to the semantics of spatial expressions. In order to test further and validate our representations, we have studied some of the inferences allowed by our first-order logical representations and compared them to the conclusions that would intuitively be available to human speakers. The methodology we have exemplified here has been followed by various members of our research group, focusing on other prepositions (Aurnague, 1995; Vieu, 1991), localization nouns (Aurnague, 1991), and motion verbs (Asher and Sablayrolles, 1995; Muller and Sarda, 1999).

9

Change of Orientation

HEDDA SCHMIDTKE, LADINA TSCHANDER,
CAROLA ESCHENBACH, and
CHRISTOPHER HABEL

Abstract

Concepts of direction play a crucial role for different groups of spatial verbs and spatial adverbs. This chapter focuses on some German verbs that encode change of orientation (*abbiegen* 'turn off', *wenden* 'turn around'). These verbs are compared to verbs of motion and verbs of orientation. With the first group they share reference to situations of spatial change; with the second group they share reference to directions that are not exclusively constituted by change of location.

To represent change of orientation, in the context of change of location, it is necessary to specify sameness of orientation and difference of orientation independently of location. This article provides a geometric description of directions that can serve as a basis for defining sameness of orientation independent of location. For this, we define half-lines, which can serve as geometric models for axes in spatial reference systems. Half-lines are instances of exact directions at a specific location. Ranges of direction are specified and applied to describe the semantic contribution of the German adverbial phrases *nach rechts, nach links, schräg, scharf* ('to the right', 'to the left', 'diagonally', 'sharply') in combination with verbs of change of orientation.

Introduction

Several verbs and adverbs encode information about spatial directions. The linguistic and conceptual embedding of this information differs, although, on

The research reported in this article was carried out in the context of the project Axiomatik räumlicher Konzepte (HA 1237/7), which is part of the priority program on Spatial Cognition of the Deutsche Forschungsgemeinschaft (DFG). The work of C. Eschenbach was also supported by the Institute of Advanced Study in Berlin.

the geometric level, there seems to be a uniform notion of spatial direction. The goal of this article is to contribute to the understanding of the different roles language assigns to directions and to the geometric background of modeling a uniform spatial notion of direction.

A short overview of some groups of German adverbs provides a first impression of the diversity of **direction-based concepts** in natural language.

- **Projective prepositions** such as *vor* ('in front of') and **projective adverbs** such as *rechts* ('right') designate the location of an object based on a direction specified by **a spatial reference system**.
- **Directional adverbial phrases** like *in das Haus, nach rechts* ('into the house', 'to the right')[1] specify directions on the basis of locations. For example, the path described by *Maria geht in das Haus* ('Mary walks into the house') ends in the interior of the house.
- The **rotational adverbs** *rechtsherum, linksherum* ('clockwise', 'counterclockwise') specify direction on a cyclic trajectory (*Das Karussell$_i$ dreht sich$_i$ rechtsherum* 'The carousel rotates clockwise') (cf. Habel, 1999).
- The **orientational adverbs** *vorwärts, rückwärts, seitwärts* ('forwards', 'backwards', 'sideways') specify the direction of motion relative to a spatial reference system (*Paul geht rückwärts* 'Paul walks backwards').[2]
- In addition, the adverbs *schräg* and *scharf* ('diagonally' and 'sharply') can be used in certain contexts to specify **changes of orientation** (*Der Wagen biegt schräg ab* 'The car turns off diagonally').[3]

There are also various spatial verbs that encode direction. The type of the verb determines how the information given by a spatial adverbial phrase is conceptually embedded.

---

[1] The German prepositions *in, an, auf, vor, hinter, über, unter* ('in(to)', 'on(to)', 'on(to)', 'in front of', 'behind', 'over', 'under') form local prepositional phrases when combined with a dative noun phrase, and directional (goal) prepositional phrases when combined with an accusative noun phrase. *In dem Haus* ('in the house') specifies the region inside the house, whereas *in das Haus* ('into the house) specifies the path leading to this region.

[2] While the technical terms 'projective adverbial phrase' and 'directional adverbial phrase' are common in the linguistic discussion, the terms 'rotational adverb' and 'orientational adverb' do not belong to the established terminology.

[3] Note that the German examples are translated literally. In all examples with *abbiegen*, we use the English expression *turn off*, which reflects the morphological complexity of the German verb. In addition the particle separates this use from other meanings of *turn*, which correspond to other verbs in German (for example *turn around* corresponds to *wenden*). *Turn* without a particle is used for the translation of *drehen*. We translate the reflexive pronoun in *sich$_i$ wenden, sich$_i$ drehen* as REFL$_i$, to avoid the idiomatic expression *turn oneself into something*. It should be clear that the English translations do not replicate all acceptability ratings of the German expressions.

- Verbs of motion such as *gehen, laufen* ('walk', 'run') express the **change of location** of an object. Directional adverbial phrases specify the PATH of the object through space.
- Verbs of orientation such as *zielen, zeigen, starren, schauen* ('aim', 'point', 'stare', 'look') can express static situations in which an object is oriented according to a direction that is specified by a directional adverbial phrase.
- Verbs of change of orientation such as *abbiegen, wenden, drehen* ('turn off', 'turn around', 'turn') combine spatial change and object orientation. Directional adverbial phrases can specify the PATH (*in die Gasse abbiegen* 'turn off into the lane') or a final **orientation** (*nach Norden drehen* 'turn to the north').

Verbs of motion and verbs of change of orientation describe the spatial change of one entity against a stable background. In *Paul walks to the tree* and *Paul turns to the tree*, Paul is presented as the object that changes relative to the background to which the tree belongs. The object to which the verb ascribes the change will be called **bearer of change** in the following. If the verb expresses change of location, then we also use the term **bearer of motion** (cf. Eschenbach *et al.*, 2000). The combinations of verbs and adverbs will be discussed in more detail below.

The types of spatial change we are interested in here are change of location and change of orientation. Figure 9.1 displays some cases. (a) A car moves straight ahead, changing location without changing orientation. (b) A trolley can be pushed forwards and sideways, thus also changing location and direction of motion without changing orientation. (c) A San Francisco cable car changes orientation—but not location—when it is rotated on a turntable. (d) A car turning off a highway changes location, orientation, and direction of motion.

The concepts of **direction of motion** and **object orientation** play a major role in the analysis of the spatial inventory of natural languages (cf. Herskovits, 1986;

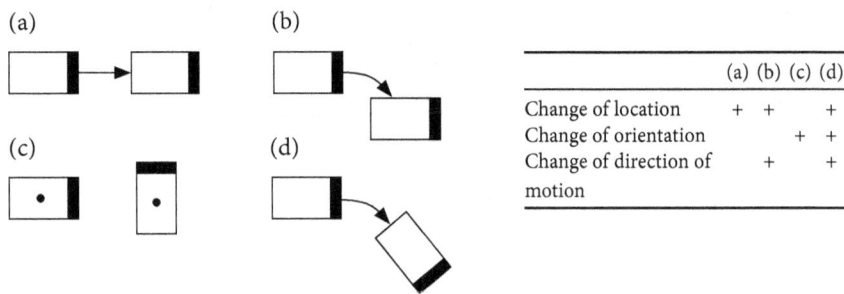

	(a)	(b)	(c)	(d)
Change of location	+	+		+
Change of orientation			+	+
Change of direction of motion		+		+

FIG. 9.1 Examples of spatial change.

Aurnague, 1995; Logan and Sadler, 1996; Carlson, 1999; Gambarotto and Muller, Ch. 8). The contrast between direction of motion and object orientation is also useful for explaining differences in the combinations of verbs and adverbs that specify directions. Since the differentiation is linguistically relevant, we assume that direction of motion and object orientation are distinguished in the lexical entries of the verbs. The geometrical specification of both concepts, however, is based on a uniform **geometric model of directions**. The goal here is to elucidate the constraints of combining verbs and adverbs on both the linguistic and the geometric level.

The next section introduces the geometric structure of **half-lines**. It is used to specify the geometric structure of spatial reference systems and the **geometric model of sameness of orientation**. With this background, the examples from German are discussed. Information about direction of motion and object orientation can play a role in the conceptual structure of both verbs and adverbs. Therefore, it is necessary to clarify what these parts of speech contribute and how they interact when they are combined. We focus on German verbs of change of orientation and compare them with of verbs of motion and verbs of orientation. The pattern of possible combinations of verbs and adverbs is ascribed to their semantic contributions. The cases of combining verbs of change of orientation with the German adverbial phrases *nach rechts*, *nach links*, *schräg*, and *scharf* ('to the right', 'to the left', 'diagonally', 'sharply') are then discussed in relation to the geometric model. To describe the contribution of these adverbials in a general manner, **ranges of direction** are finally specified on the geometric level.

1. Object Orientation, Spatial Reference Systems, and Directions

We describe the concept of direction with an **axiomatic system**. An axiomatic characterization of a system of concepts specifies semantic interrelations between and semantic properties of the concepts. The specification is based on half-lines, which represent a geometric model of **direction** that is neutral regarding **length** or **distance**. This reflects the fact that in descriptions of spatial constellations information about lengths or distances is optional. This procedure provides the opportunity to separate investigations of the concepts of direction and distance. Nevertheless, metrical concepts can be added to the framework if the specification of distance or length is demanded for some reason. The approach is also open to the addition of a **geometric model of bounded directions** (called '**vectors**' by Zwarts, 1997).

Half-lines can represent the geometric structure of the **axes of a spatial reference system**. We start this section with a short discussion of the role and geometric structure of spatial reference systems. In the second part of this section, we describe the geometry of **points**, **straight lines**, and half-lines. These concepts are used to express the geometric constraints on spatial reference systems. The relation of **equidirection** is then defined. It is the core concept of the geometry of **exact directions** and ranges of direction and used for analyzing verbs of orientation and verbs of change of orientation in section 2.

Every geometric structure can be formalized in a number of ways. Therefore, we do not claim that half-lines are necessarily mentally represented as described here. Rather, we show that a geometric framework based on points, straight lines, and half-lines provides a sound basis for characterizing the geometric nature of the linguistic concepts of direction of motion and change of orientation.

1.1. Object Orientation and Spatial Reference Systems

An object can be oriented only if it has a **distinguished axis**. For example, in describing a brick as being vertically or horizontally oriented, we refer implicitly to the **maximal axis** of a rectangular parallelepiped, whereas a cube—since it does not have a maximal axis—cannot be described as being vertically or horizontally oriented (Lang, 1990). Spatial reference systems, which are the fundamental conceptual structures for the usage of spatial concepts in language and cognition, are often characterized as systems of axes. These axes can be determined by functional properties of objects or by the direction of motion of an object (Jackendoff and Landau, 1993). Herskovits (1986) argues that spatial reference systems are based on six half axes originating at a central location, rather than three axes running through the central location.

Spatial reference systems are conceptual structures that can involve functional aspects (Carlson, 1999; Gambarotto and Muller, 2000). Their geometric models are called '**frames of reference**' (Eschenbach, 1999). Several researchers follow the mathematical tradition of Cartesian geometry and describe frames of reference as **coordinate systems**. However, alternative geometric models have been successfully used, for example in the analysis of projective prepositions (Eschenbach and Kulik, 1997; Eschenbach, 1999). In this chapter we combine the latter approach with the proposal of Herskovits (1986). Half-lines form the geometric specification of axes that originate at a location, and their specification is independent of metrics and coordinates.

A perceptual counterpart to a half-line is a person's **line of sight**, as depicted in Fig. 9.2. We exemplify this with the case of Mary looking towards the sea. Mary's eyes define a starting point for a half-line f that runs across the water. With respect to her orientation as well as to what she sees, it is irrelevant what

Fig. 9.2 Half-lines induced by the line of sight.

is at Mary's back. Nevertheless, to answer the question, What is behind Mary?, a second half-line *b*, which is the **converse** of *f*, is needed.

To simplify the following presentation of geometric structure, we employ a **planar geometry**. Therefore, frames of reference (characterized in sect. 1.3) consist of four half-lines (representing the front axis, the back axis, the right axis, and the left axis, respectively).

To determine whether two objects have the same orientation, the half-lines representing their characteristic axes have to be compared independently of the location of the objects. Therefore, **sameness of direction** should be described independently of a location. A half-line starts at some point and has a characteristic direction. So, half-lines can be seen as **instantiated directions**, i.e. directions rooted at a point. On the other hand, directions represent the characteristics of half-lines that are independent of their location or starting point. Thus, directions and half-lines provide a basis for specifying the geometric aspects of concepts like direction of motion and object orientation.

1.2. The Geometric Framework: Axioms for Points, Straight Lines, and Half-Lines

1.2.1. *Axioms for Points and Straight Lines*

Geometric directions are characterized on the basis of the **incidence geometry** for straight lines, half-lines, and points. The axioms of incidence specify the relationship between straight lines and points. A point P is **incident** (ι) with a straight line *g* if it lies on the line. In this case, we will also say that the line contains the point. The first two axioms specify that every straight line has at least two points (I1)[4] and that for every pair of points there is some straight line on which both of them lie (I2). Axiom (I3) says that if two points are incident with the straight lines g_1 and g_2, then g_1 and g_2 are identical. This corresponds to the statement that different straight lines do not share more than one point. Finally, no straight line contains all points (I4).

---

[4] The labels of the axioms start with a letter identifying the geometric relation or geometric entity that is described with the axioms (as incidence (I), parallelity (P), half-lines (H), reference system (R) and equidirection (E)). Definitions are marked with (D), and specifications of lexemes are labeled with (L).

(I1) $\forall g: \exists P, Q: P \neq Q \wedge P \iota g \wedge Q \iota g$
(I2) $\forall P, Q: \exists g: P \iota g \wedge Q \iota g$
(I3) $\forall P, Q, g_1, g_2: P \neq Q \wedge P \iota g_1 \wedge P \iota g_2 \wedge Q \iota g_1 \wedge Q \iota g_2 \rightarrow g_1 = g_2$
(I4) $\forall g: \exists P: \neg(P \iota g)$

One condition for having the same orientation is that the two half-lines involved are part of the same straight line or of parallel straight lines. To combine these two conditions, we employ the reflexive variant of the notion of **parallelity** (in contrast to Euclid): Two straight lines are parallel iff they are identical or do not share any point (D1). The **transitivity of parallelity** is guaranteed by (P1). Consequently, the relation of parallelity is an **equivalence relation**. Axiom (P2) then says that, given a point P and a straight line g, there is a parallel straight line g' through P. These axioms ensure that our geometry is a Euclidean plane with respect to parallelity.

(D1) $g_1 \parallel g_2 \Leftrightarrow_{\text{def}} g_1 = g_2 \vee \neg \exists P: P \iota g_1 \wedge P \iota g_2$
(P1) $\forall g_1, g_2, g_3: g_1 \parallel g_2 \wedge g_2 \parallel g_3 \rightarrow g_1 \parallel g_3$
(P2) $\forall g, P: \exists g': P \iota g' \wedge g \parallel g'$

1.2.2. Axioms for Half-Lines

We restrict our presentation of half-lines to the axioms that are necessary to define directions. Schmidtke (1999) presents a more thorough investigation of a geometry based on half-lines. The idea of splitting a straight line into two half-lines is captured by the assumptions that half-lines are parts of lines (H1), that every half-line has a starting point (H2 and H4), and that for every half-line there is a converse half-line (H6). The converse half-line contains the points of the straight line that do not lie on the given half-line.[5]

We begin with the definition of the relation **part of** ($\subseteq$) as a convenient abbreviation for all kinds of geometric entities. A geometric entity x is part of a geometric entity y if all points that lie on x also lie on y (D2). Every half-line is part of a straight line (H1). Even though straight lines and half-lines are not sets of points, their points uniquely identify them. Two half-lines are identical if they have the same points (H2). This makes the relation part of **anti-symmetric** on the set of half-lines. In addition, we will speak of **parallel half-lines** ($\parallel$) if they are parts of parallel straight lines (D3), illustrated in Fig. 9.3b.

---

[5] Since half-lines are parts of straight lines, we obtain a basis to say that half-lines are straight. **Oriented curves**, which can represent paths of motion (cf. Eschenbach, Habel, and Kulik, 1999; Eschenbach et al., 2000), in contrast, need not be straight. An oriented curve is a linear, bounded structure with an intrinsic direction, which orders the locations on it and distinguishes starting point and final point. However, this aspect will not be focused on in the following.

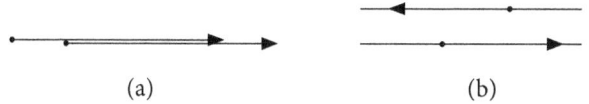

FIG. 9.3 (a) A half-line that is part of another half-line. (b) Parallel half-lines.

(D2) $x \subseteq y \Leftrightarrow_{\text{def}} \forall P: P \iota x \rightarrow P \iota y$
(H1) $\forall r \exists g: r \subseteq g$
(H2) $\forall r, r': [\forall P: P \iota r \leftrightarrow P \iota r'] \leftrightarrow r = r'$
(D3) $r_1 \parallel r_2 \Leftrightarrow_{\text{def}} \exists g_1, g_2: r_1 \subseteq g_1 \wedge r_2 \subseteq g_2 \wedge g_1 \parallel g_2$

The relations part of and parallel are fundamental for relating half-lines. But what we need in addition is the possibility of comparing the points on a half-line with respect to the question of which point lies further away in the direction of the half-line. The first step is to characterize the starting point of a half-line. A starting point of r lies on r and does not lie on any other half-line that is part of r (D4). Every half-line has a unique starting point (H3) and contains other points in addition (H4). Given a point P on a straight line g, there is a half-line r that is part of g and starts in P (H5). Since every half-line has a starting point and the starting point of a half-line is unique, we can use 'start' also to denote the function that maps a half-line to its starting point (D5).

(D4) $\text{start}(r, P) \Leftrightarrow_{\text{def}} P \iota r \wedge [\forall r': r' \subseteq r \wedge P \iota r' \rightarrow r' = r]$
(H3) $\forall r \exists P: \text{start}(r, P) \wedge \forall Q: \text{start}(r, Q) \rightarrow Q = P$
(H4) $\forall r \exists P: \neg \text{start}(r, P) \wedge P \iota r$
(H5) $\forall g, P: P \iota g \rightarrow \exists r: r \subseteq g \wedge \text{start}(r, P)$
(D5) $P = \text{start}(r) \Leftrightarrow_{\text{def}} \text{start}(r, P)$

The converse of a half-line r is a half-line that lies on the same straight line as r, shares nothing but the starting point with r, and contains all points of the straight line that are not on r (D6). Every half-line has a converse (H6).

(D6) $r' = \text{conv}(r) \Leftrightarrow_{\text{def}} \forall P: P \iota r' \leftrightarrow (\text{start}(r, P) \vee \exists g: r \subseteq g \wedge P \iota g \wedge \neg (P \iota r))$
(H6) $\forall r \exists r': r' = \text{conv}(r)$

The following axiom forms the basis for specifying the ordering relation for the points on a half-line. If two half-lines r_1 and r_2 are part of a common straight line and the starting point of r_1 lies on r_2, but the starting point of r_2 does not lie on r_1, then r_1 is part of r_2 (H7). This configuration is depicted in Fig. 9.4a. Figs. 9.4b and c display the other configurations of half-lines that are part of a common straight line.

(H7) $\forall r_1, r_2: [\exists g: r_1 \subseteq g \wedge r_2 \subseteq g] \wedge \text{start}(r_1) \iota r_2 \wedge \neg (\text{start}(r_2) \iota r_1) \rightarrow (r_1 \subseteq r_2)$

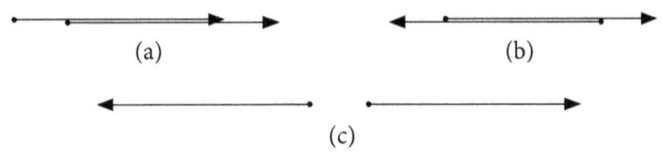

FIG. 9.4 Configurations of two half-lines that are part of the same straight line.

We can visualize the situation covered by (H7) with reference to the lines of sight of two people, Mary and Paul. If Mary sees Paul and there is something else they both see, their configuration is either that of Fig. 9.4a or that of Fig. 9.4b. In Fig. 9.4a they both look in the same direction, so Mary can see everything that Paul sees (if Paul is not too tall). In Fig. 9.4b they face each other, so Mary will see what is at Paul's back. If Mary and Paul turn their backs to each other, then they do not see any common region, as in Fig. 9.4c.

The ordering of points on a half-line is made explicit by defining the ternary relation of **precedence** (<). P precedes Q on r iff P is different from Q, and a half-line r' that is part of r starts in P and contains Q (D7). Note that $P <_r Q$ is another way of writing $<(r, P, Q)$.

(D7) $P <_r Q \Leftrightarrow_{\text{def}} P \neq Q \wedge \exists r': r' \subseteq r \wedge \text{start}(r', P) \wedge Q \iota r'$

It can be shown that the characterization of half-lines given above suffices to order the points of r by the relation $<_r$. All the half-lines that are part of a straight line can be divided into two groups that provide compatible orderings. For each half-line in one group, its converse is in the other group. Thus we can specify which of two points that lie on the same half-line is farther in the direction of the half-line. In addition, we know that a half-line and its converse provide different directions (cf. Schmidtke, 1999).

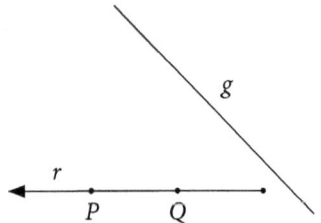

FIG. 9.5 P and Q lie on the same side of g.

For the definition of **equidirection** (given in sect. 1.4), the relation of being on the same side (σ) of a straight line that relates points and straight lines in the plane is needed. Two points P and Q are **on the same side** of a straight line g (D8) iff they lie on a half-line r that does not share any point with g.

(D8) $\sigma(g, P, Q) \Leftrightarrow_{def} \exists r: P \iota r \wedge Q \iota r \wedge \neg \exists R: R \iota g \wedge R \iota r$

We are now able to specify the relevant geometric constraints for frames of reference and the central relation for the description of object orientation and change of orientation.

1.3. Geometric Constraints on Frames of Reference

Frames of reference are the geometric structures that represent spatial reference systems. We use 'front-axis(SRS(x, t))' to refer to the half-line that represents the **front axis** of a spatial reference system that is provided by an object x at time t. The symbols **back-axis**, **right-axis**, and **left-axis** are used accordingly. The starting point of front-axis(SRS(x, t)) is the location of x at t.[6]

The geometric constraints on spatial reference systems (*srs*) are the following. The back axis is always converse to the front axis of the same spatial reference system (R1). The right axis is not parallel to the front axis (or the back axis) (R2),[7] and the left axis is converse to the right axis (R3).

(R1) back-axis(*srs*) = conv(front-axis(*srs*))
(R2) ¬(right-axis(*srs*) ∥ front-axis(*srs*))
(R3) left-axis(*srs*) = conv(right-axis(*srs*))

1.4. Object Orientation and the Relation of Equidirection

The concept of change of orientation is based on a comparison of the initial and the final orientation of the bearer of change. This comparison should be possible even if the change involves change of location (cf. Figs. 9.1a, b, d above). The orientation of an object in the plane depends on its front axis. Two objects have the **same orientation** if the half-lines representing their front axis have the same direction (Fig. 9.6a). Two objects have **opposite orientations** if these half-lines

---

[6] Please note that the half-line representing the front axis is derived from the spatial reference system SRS(x, t), not from the object x at time t itself. For constructing spatial reference systems not depending on an object—e.g. the **absolute reference system** for cardinal directions (cf. Schmidtke, 2001)—other parameters may be necessary. However, the construction of spatial reference systems is not the focus of this chapter.

[7] (R2) expresses mutual independence of the right and the front axis. However, the axes do not have to be orthogonal to fulfill their function.

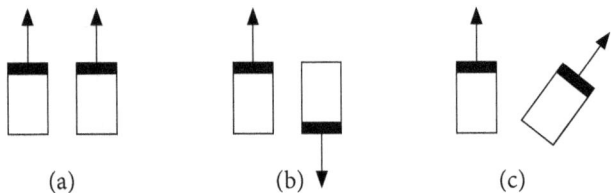

FIG. 9.6 (a) same orientation; (b) opposite orientation; (c) different orientation.

have converse directions (Fig. 9.6b). If the half-lines are not parallel, then the two objects have different (but not opposite) orientations (Fig. 9.6c).

The sameness of orientation is geometrically modeled by the relation of **equidirection** (symbolized as ↑↑) applied to the half-lines that represent the front axes of the object(s). Before we give a definition for this relation, we outline the general principles that should be fulfilled by such a relation.

The first group of assumptions says that equidirection is an equivalence relation, i.e. symmetric (E1), transitive (E2), and reflexive (E3).

(E1) $\forall r_1, r_2: r_1 \uparrow\uparrow r_2 \rightarrow r_2 \uparrow\uparrow r_1$
(E2) $\forall r_1, r_2, r_3: r_1 \uparrow\uparrow r_2 \wedge r_2 \uparrow\uparrow r_3 \rightarrow r_1 \uparrow\uparrow r_3$
(E3) $\forall r: r \uparrow\uparrow r$

Thus, if we know what sameness of direction is, we can obtain directions as entities: **exact directions** (xd = eXact Direction) are the equivalence classes of equidirection (D9). Exact directions are used in sect. 2.4 for characterizing complex verb phrases.

(D9) $r \in \text{xd}(r') \Leftrightarrow_{\text{def}} r \uparrow\uparrow r'$

Directions as defined here abstract from the starting points of the underlying half-lines. This is in analogy to the geometric definition of a **vector** as an equivalence class of all arrows that have the same length and the same direction. If we would use arrows as the basic geometric entities, then directions could be understood as abstracting from both the starting point and the length of the underlying arrows.

In addition to this general consideration, substantial constraints derive from the geometric specification of half-lines. **Equidirected half-lines** are parallel (E4), and a half-line and its parts are equidirected (E5). A half-line and its converse are **not equidirected** (E6), but the **converse half-lines** of two equidirected half-lines are (E7).

(E4) $\forall r_1, r_2: r_1 \uparrow\uparrow r_2 \rightarrow r_1 \parallel r_2$

(E5) $\forall r_1, r_2: r_1 \subseteq r_2 \to r_1 \uparrow\uparrow r_2$
(E6) $\forall r: \neg (r \uparrow\uparrow \mathrm{conv}(r))$
(E7) $\forall r_1, r_2: r_1 \uparrow\uparrow r_2 \to \mathrm{conv}(r_1) \uparrow\uparrow \mathrm{conv}(r_2)$

Furthermore, we assume that every direction can be instantiated at every point. This means that for every point P and every half-line r, there is a unique half-line r′ that is equidirected to r and starts in P (E8).

(E8) $\forall r, P: \exists r': r' \uparrow\uparrow r \wedge \mathrm{start}(r', P) \wedge \forall r'': \mathrm{start}(r'', P) \wedge r'' \uparrow\uparrow r \to r'' = r'$

All conditions for equidirection are fulfilled if equidirection is defined as in (D10). Two half-lines are **equidirected** iff they are parallel and there is a straight line g that is not parallel to them such that all points of the half-lines lie on the same side of g. This is illustrated in Fig. 9.7.

(D10) $r_1 \uparrow\uparrow r_2 \Leftrightarrow_{\mathrm{def}} r_1 \| r_2 \wedge \exists g: \neg(g \| r_1) \wedge \forall P, Q: P \iota r_1 \wedge Q \iota r_2 \to \sigma(g, P, Q)$

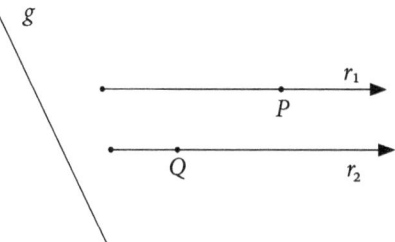

FIG. 9.7 Equidirected half-lines.

Now we return to the examples of Fig. 9.1. In the cases depicted by Figs. 9.1c and d a change of orientation takes place. The half-lines that represent the front axes at the beginning (t_1) and at the end of the movement (t_2), respectively, are not equidirected. In the case depicted by Figs. 9.1a and b, a change of orientation has not taken place. Correspondingly, the half-lines that represent the front axes at the beginning and at the end of the movement are equidirected. This relation of equidirection is used in the next section for analyzing verbs of change of orientation.

2. Verbs of Change of Orientation: Linguistic Characterization and Geometric Specification

Verbs of change of orientation are generally not covered by the analysis of verbs of motion (cf. Jackendoff, 1990; Maienborn, 1990; Kaufmann, 1995;

Ehrich, 1996; Eschenbach *et al.*, 2000). The common feature of verbs of change of orientation and verbs of motion is that both express spatial change. However, verbs of change of orientation also share some characteristics with (static) verbs of orientation. In the next section, we present the semantic components of verbs of motion and verbs of orientation. After that, we look more closely at the components that are relevant for the semantics of verbs of change of orientation. The focus of the linguistic discussion of the verb groups is on their ability to combine with spatial adverbial phrases. We assume that some restrictions derive from the semantic structure of the verbs and the adverbial phrases and that other restrictions derive from geometric constraints.

2.1. Verbs of Motion and Verbs of Orientation

Verbs of motion (*gehen, schieben, eintreten, stellen* ('walk', 'push', 'enter', 'put')) express a change of location. The semantic component characteristic of verbs of motion is $GO(x, w)$, which encodes that the bearer of motion x moves along the **path of motion** w. This component is also responsible for relating the path of motion to time.[8] Verbs that focus on the final location (*eintreten* 'enter') contain the additional semantic component $BE\ AT(x, p)$, which specifies the final location p of the bearer of motion. Specific semantic components add constraints on the mode of motion (for example the idiosyncratic difference between *walk, run, stroll*) (cf. Eschenbach *et al.*, 2000).

The parameter for the path of motion in the semantic representation of the verb of motion is the anchor for adding specific information about the direction of motion. This can be done by directional adverbial phrases, which express purely geometric conditions. The lexemes of this class systematically disregard the orientation of the bearer of motion relative to the PATH. Correspondingly, the formal characterization of the semantic component $GO(x, w)$ takes no account of the orientation of the bearer of motion.

The combination with the orientational adverb *rückwärts* ('backwards'), as in (1), shows that information on the orientation of the bearer of motion relative to the path of motion can be supplied. The semantics of the verb of motion is compatible with the semantics of the orientational adverb. However, the information about the orientation of the bearer of motion is optional. The regular connection between the direction of motion and the orientation of the bearer of motion is that they are identical by default. Thus, motion is forward, if not stated

---

[8] All information about the temporal structure of the event or process in question is included in the semantic component of the verbs, whereas the directional adverbial phrase only gives information about spatial conditions.

otherwise. Taking this condition as a general pragmatic default, we do not need to assume that it is explicitly specified by the semantics of the verbs.

Rotational adverbs such as *rechtsherum* ('clockwise') cannot be combined with verbs of motion. The sentence (2) might be interpreted as specifying a path of motion around an obstacle, but this is valid only in very specific contexts.[9]

(1) Paul geht rückwärts. ('Paul walks backwards.')
(2) *Paul geht rechtsherum. ('Paul walks clockwise')

Verbs of orientation (*zielen, zeigen, starren, schauen* ('aim', 'point', 'stare', 'look')) specify an axis that can also be specified by directional adverbials (3). Thus, we assume that the characteristic semantic component for verbs of orientation is v-axis(x) ↑↑ w. It expresses the fact that the axis of the object x that is specified by the verb (v-axis) is equidirected with the PATH w. This path is called **path of orientation** in what follows.

(3) Paul zielt/zeigt/starrt/schaut in das Haus. ('Paul aims/points/stares/ looks into the house.')

Verbs of orientation do not express change. Therefore they do not contain the component GO(x, w). This corresponds to the observation that verbs of orientation do not combine with orientational adverbs *rückwärts* ('backwards') as in (4) and (5), which shows that orientational adverbials depend on the existence of a path of motion in the verb meaning.[10] Similar to verbs of motion, verbs of orientation do not combine with rotational adverbs (6).

(4) *Paul zielt/starrt/schaut rückwärts. ('Paul aims/stares/looks backwards.')
(5) *Der Pfeil zeigt rückwärts. ('The arrow points backwards.')
(6) *Paul zielt/zeigt/starrt/schaut rechtsherum. ('Paul aims/points/ stares/looks clockwise.')

To sum up, directional, orientational, and rotational adverbial phrases are sensitive to different semantic components in the lexical representation of verbs. Directional adverbials and orientational adverbs call for a PATH w in the

---

[9] Here and in the following, an asterisk (*) preceding an expression signals that the expression is not acceptable or does not lead to a clear interpretation. If the judgements about a sentence are ambiguous and vary between speakers or contexts of use, we use '?'.

[10] *Zeigen* ('point') with a human subject can be combined with orientational adverbs (?*Paul zeigt rückwärts* 'Paul points backwards'). This means either that this verb can denote the movement of the hand or that the dissociation of the axis of pointing and the front axis of humans is the critical factor.

semantic representation. In the semantic representation of the verb, the PATH w is embedded either by the motion component ($GO(x, w)$) or by the orientation component (v-axis(x) ↑↑ w). Orientational adverbs require that the object provides an **intrinsic reference system** that is not fixed by the verb meaning (denoted with IRS(x)). In the rightmost row of Table 9.1, the sensitivities of the adverbial phrases are displayed. The cell for rotational adverbs is left free, since we know only that the availability of paths is not sufficient for their use. Table 9.1 also lists the characteristics of verbs of motion and verbs of orientation. Since verbs of motion specify paths of motion, they can be combined with different spatial adverbials. Since verbs of orientation do not refer to change, they can combine only with directional adverbials that can describe static orientation.

TABLE 9.1 *Verbs of motion and verbs of orientation combined with spatial adverbial phrases*

	Verb of motion	Verb of orientation	Requirement for adverbial phrase
Directional adverbial phrase	Paul geht in die Gasse.	Paul starrt in die Gasse.	PATH w
	'Paul walks into the lane.'	'Paul stares into the lane.'	
Orientational adverb	Paul geht rückwärts.	*Paul starrt rückwärts.	$GO(x, w)$ and IRS(x)
	'Paul walks backwards.'	'Paul stares backwards.'	
Rotational adverb	*Paul geht rechtsherum.	*Paul starrt rechtsherum.	
	'Paul walks clockwise.'	'Paul stares clockwise.'	
Semantic component of verb	$GO(x, w)$	v-axis(x) ↑↑ w	

2.2. Comparison of Verbs of Change of Orientation with Verbs of Motion and Verbs of Orientation

Verbs of change of orientation (*abbiegen* ('turn off'), *wenden* ('turn around'), *drehen* ('turn')) explicitly indicate a change of the orientation of the bearer of motion. Sentence (7), for example, can be applied only to a situation if Paul's orientation changes. Correspondingly, an adverb that indicates that the object orientation remains the same during the movement cannot be combined with *abbiegen* (8).

(7) Paul biegt ab. ('Paul turns off.')
(8) *Paul biegt geradeaus ab. ('Paul turns off straight ahead.')

As mentioned above, an object can change its orientation without changing location. The German verb *drehen* is neutral regarding change of location (cf. Habel, 1999). It usually applies to rotational movements that leave the bearer of change in place (as *turn the knob to the right*). On the other hand, the German verb *abbiegen* expresses the combination of a change of location and a change of orientation of one entity. Correspondingly, *abbiegen* can be combined with orientational adverbs, but *drehen* cannot.

(9) Paul biegt rückwärts ab. ('Paul turns off backwards.')
(10) *Paul dreht den Knopf rückwärts. ('Paul turns the knob backwards.')

The German verb *wenden* ('turn around') can apply to both kinds of change of orientation. Sentence (11) means that the paper stays in place, while it is evidently impossible to turn the car without changing location in (12). The addition of an orientational adverb is more acceptable, if a path of motion can be inferred (13), (14). If the verb *wenden* is not combined with a directional adverbial phrase, it expresses that the final orientation is opposite to the initial orientation of the bearer of change.

(11) Paul wendet das Papier. ('Paul turns around the paper.')
(12) Paul wendet das Auto. ('Paul turns around the car.')
(13) *Paul wendet das Papier rückwärts. ('Paul turns around the paper backwards.')
(14) ?Paul wendet das Auto rückwärts. ('Paul turns around the car backwards.')

The combination of verbs of change of orientation and directional adverbials shows that we have to distinguish two types of goal-directed adverbial phrases. The verbs of change of orientation can be combined with *nach rechts* (15)–(18). However, only *abbiegen* can be combined with the directional adverbial phrase *in die Gasse* (19)–(22). The acceptability of (21) is disputable.

(15) Paul$_i$ dreht sich$_i$ nach rechts. ('Paul$_i$ turns REFL$_i$ to the right.')
(16) Paul$_i$ wendet sich$_i$ nach rechts. ('Paul$_i$ turns REFL$_i$ to the right.')

(17) ?Paul wendet das Auto nach rechts. ('Paul turns the car to the right.')
(18) Paul biegt nach rechts ab. ('Paul turns off to the right.')
(19) *Paul$_i$ dreht sich$_i$ in die Gasse. ('Paul$_i$ turns REFL$_i$ into the lane.')
(20) *Paul$_i$ wendet sich$_i$ in die Gasse. ('Paul$_i$ turns REFL$_i$ into the lane.')
(21) ?Paul wendet das Auto in die Gasse. ('Paul turns the car into the lane.')
(22) Paul biegt in die Gasse ab. ('Paul turns off into the lane.')

In (15)–(18), *nach rechts* specifies the final orientation of the bearer of change. In (16) and (17) *in die Gasse* specifies the path of motion. The restricted acceptability of (14) and (21) indicates that the path of motion is not a constitutive element in the semantics of *wenden* but might be associated on demand (Maienborn, 1994). *In die Gasse* differs from *nach rechts* in that the goal provides a limit. *Drehen* cannot be combined with directional adverbial phrases whose goals provide a limit.[11] Correspondingly, we will restrict the analysis of the combination of directional adverbial phrases with verbs of change of orientation in the next section to *nach rechts* and *nach links*. Verbs of orientation can be combined with both kinds of goal-directed adverbial phrases. This is the reason why we assumed above that they relate the axis to a path of orientation rather than to another axis.

Rotational adverbs can be combined with verbs of change of orientation (23)–(25). However, this combination is less acceptable for *abbiegen*.

(23) Paul$_i$ dreht sich$_i$ rechtsherum. ('Paul$_i$ turns REFL$_i$ clockwise.')
(24) Paul$_i$ wendet sich$_i$ rechtsherum. ('Paul$_i$ turns REFL$_i$ clockwise.')
(25) Paul wendet das Auto rechtsherum. ('Paul turns the car around clockwise.')
(26) ?Paul biegt rechtsherum ab. ('Paul turns off clockwise.')

Table 9.2 summarizes the characteristic semantic components for verbs of change of orientation. We use the component CH-ORIENT(x) to represent the change of orientation of x.

The next section is devoted to the question of how to describe the geometric conditions of change of orientation and the role of the spatial reference system. The combination of the verbs of change of orientation with the adverbial phrases *nach rechts*, *nach links*, *scharf*, and *schräg* ('to the right', 'to the left', 'sharply', 'diagonally') should be derivable from this analysis in a systematic manner. Habel (1999) presents an analysis of the rotational adverbs combined with *drehen*.

---

[11] A remarkable consequence is that, for adverbial phrases formed with the preposition *zu* ('towards'), the goal does not function as a limit (cf. Kaufmann, 1995).

TABLE 9.2 *Verbs of change of orientation combined with spatial adverbial phrases*

	Verb of change of orientation	Verb of change of orientation and location	Requirement of the adverbial phrase
Directional adverbial phrase (bounded)	*Paul$_i$ dreht sich$_i$ in die Gasse. 'Paul$_i$ turns REFL$_i$ into the lane.'	Paul biegt in die Gasse ab. 'Paul turns off into the lane.'	PATH w
Directional adverbial phrase (unbounded)	Paul$_i$ dreht sich$_i$ nach rechts. 'Paul$_i$ turns REFL$_i$ to the right.'	Paul biegt nach rechts ab. 'Paul turns off to the right.'	PATH w or CH-ORIENT(x)
Orientational adverb	*Paul$_i$ dreht sich$_i$ rückwärts. 'Paul$_i$ turns REFL$_i$ backwards.'	Paul biegt rückwärts ab. 'Paul turns off backwards.'	GO(x, w) and IRS(x)
Rotational adverb	Paul$_i$ dreht sich$_i$ rechtsherum. 'Paul$_i$ turns REFL$_i$ clockwise.'	?Paul biegt rechtsherum ab. 'Paul turns off clockwise.'	CH-ORIENT(x)
Semantic component of the verb	CH-ORIENT(x)	GO(x, w), CH-ORIENT(x)	

2.3. Verbs of Change of Orientation: Geometric Constraints

The goal of this section is to clarify the geometric constraints on orientation and change of orientation given by verbs and adverbial phrases.[12] We characterize *abbiegen* and *wenden* based on the geometric representation presented above. Figure 9.8 shows two cases of *abbiegen*. In fact, both cases can be described as *nach rechts abbiegen* ('turn off to the right').

The geometric characterization given in (L1) fits both cases of Fig. 9.8. (The expression 'abbiegen' (x, t_1, t_2)' and similar expressions are used here for presentational purposes only. It abbreviates the following: the verb *abbiegen* can be applied to the bearer of change x in a situation where t_1 defines the ini-

---

[12] The focus is on the spatial characteristics of the expressions. We ignore the argument structure of the verbs and do not specify complete lexical entries for the lexemes.

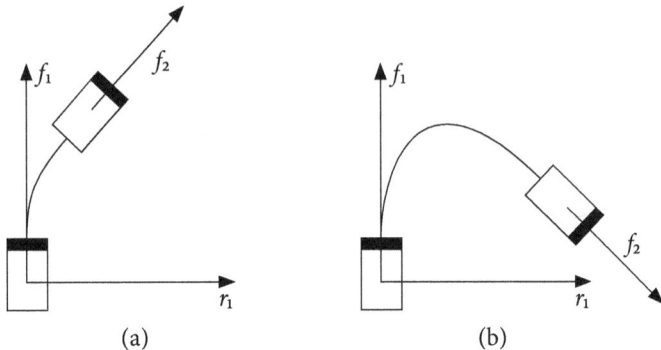

FIG. 9.8 Depictions of cases of *abbiegen* ('turn off').

tial condition and t_2 defines the final condition.) The geometric constraints require that the front axis of the bearer of change at t_2 is equidirected neither to the former front axis nor to the former back axis. This is to say that the front axis at t_2 is not parallel to the front axis at t_1. $\text{SRS}(x, t_1)$ stands for the spatial reference system contributed by the bearer of change at the beginning of the turn. *Abbiegen* seems to be neutral with respect to the question of whether the intrinsic reference system or the motion based reference system is employed (cf. Eschenbach, 1999).

(L1) $\text{abbiegen}'(x, t_1, t_2) \Rightarrow \neg(\text{front-axis}(\text{SRS}(x, t_2)) \parallel \text{front-axis}(\text{SRS}(x, t_1)))$

However, the geometric constraints for *wenden* ('turn around') are stronger. The bearer of change has to reach the opposite orientation. This condition is specified in the geometric representation (L2). Given the acceptability of (16) and (17) above, this condition seems to be too strong. However, this inference is invalid only if the final orientation is explicitly specified otherwise.

(L2) $\text{wenden}'(x, t_1, t_2) \Rightarrow \text{front-axis}(\text{SRS}(x, t_2)) \uparrow\uparrow \text{back-axis}(\text{SRS}(x, t_1))$

2.4. Final Orientations Specified by Adverbial Phrases

The scenes given in Fig. 9.8 differ with respect to the degree of bend. However, both scenes can be described by *nach rechts abbiegen* ('turn off to the right'). The adverbial phrase *nach rechts* therefore does not imply that the front axis at the final stage is equidirected to the former right axis. For the geometric characterization of complex verb phrases we therefore need to specify ranges of directions in addition to the exact directions (xd) introduced in section 1.4. **Ranges of direction** (rd) are defined on the basis of three half-lines and an order of directions (dbtw (direction between)). The geometric characterizations of these notions can be found in the next section.

In the case of Fig. 9.8b repeated here as Fig. 9.9a, the resulting front axis f_2 is included in the range between the former front axis f_1 and the former back axis conv(f_1) that includes the former right axis (r_1). The limiting axes are excluded from the range. A diagram of half-lines starting at a common point can depict this relation (Fig. 9.9b).

Direction diagrams illustrate the relations between half-lines. **Directions of half-lines** can be easily compared visually if the half-lines start at a common point. Therefore, direction diagrams display half-lines that start at a common point and are equidirected to the half-lines in question. This method is also used in vector geometry, in which vectors are often visualized by arrows starting in the origin of a coordinate system.

The directional adverbial phrase *nach rechts* specifies the range of direction in which the resulting front axis must lie. It says that the resulting direction front-axis(SRS(x, t_2)) has to be included in the range delimited by the former front axis and the former back axis and selected by the former right axis (cf. Eschenbach and Kulik, 1997; Eschenbach, 1999). The range of directions given by *nach links* is similar to that of *nach rechts*. In this case the left axis selects a range delimited by the front axis and the back axis.

(L3) nach-rechts-abbiegen'(x, t_1, t_2) $\Rightarrow$
front-axis(SRS(x, t_2)) $\in$ rd(front-axis(SRS(x, t_1)), right-axis(SRS(x, t_1)), back-axis(SRS(x, t_1)))

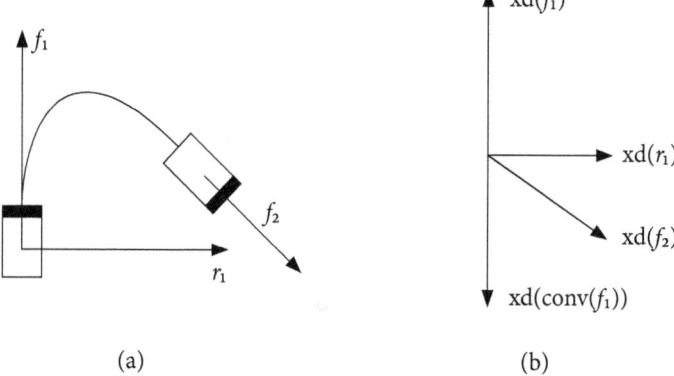

(a)   (b)

FIG. 9.9 (a) An example of *nach rechts abbiegen* ('turn to the right'). (b) The corresponding direction diagram.

(L4) nach-links-abbiegen$'(x, t_1, t_2) \Rightarrow$
front-axis(SRS$(x, t_2)) \in$ rd(front-axis(SRS(x, t_1)), left-axis(SRS(x, t_1)), back-axis(SRS(x, t_1))))

The front axis of x at t_2 is specified relative to the axes of x at t_1. Therefore we can say that x at t_1 specifies the spatial reference system on which the evaluation of the projective adverb *rechts* is based. This suggests that the verb *abbiegen* determines the spatial reference system relative to which the directional adverbial phrases are evaluated. The spatial reference system is based on the bearer of change at the beginning of the turn (srs = SRS(x, t_1)). *Abbiegen* also specifies that the front axis of the bearer of change is the axis to be considered (r = front-axis(SRS(x, t_2))). *Nach rechts* and *scharf* relate the axis r to the spatial reference system srs. (How this can be accomplished in terms of semantic composition is disregarded here and in the following.)

(L5) abbiegen$'(x, t_1, t_2) \Rightarrow$
r = front-axis$(x, t_2) \land srs$ = SRS$(x, t_1) \land \neg(r \parallel$ front-axis$(srs))$

(L6) nach-rechts$'(r, srs) \Rightarrow$
$r \in$ rd(front-axis(srs), right-axis(srs), back-axis(srs)))

(L7) nach-links$'(r, srs) \Rightarrow$
$r \in$ rd(front-axis(srs), left-axis(srs), back-axis(srs)))

Going back to the scenes of Fig. 9.8, we see that they differ with respect to the degree of change. The adverbs *schräg* ('diagonally') and *scharf* can be combined with the directional adverbial phrase or the verb *abbiegen* to specify the degree of change of orientation (27)–(29).

(27) Paul biegt scharf/schräg ab.   ('Paul turns off sharply/diagonally.')
(28) Paul biegt scharf/schräg nach rechts ab.   ('Paul turns off sharply/diagonally to the right.')
(29) Paul biegt nach scharf/schräg rechts ab.   ('Paul turns off to the sharp/diagonal right.')

Drehen cannot be combined with *scharf* or *schräg* (30). Sentences (31) and (32) demonstrate that *wenden* can only be combined with *scharf* and that this combination is possible only if a path of motion is present.

(30) *Paul$_i$ dreht sich$_i$ scharf/schräg.   ('Paul$_i$ turns REFL$_i$ sharply/diagonally.')
(31) Das Auto wendet scharf/*schräg.   ('The car makes a sharp/diagonal turn.')
(32) *Paul$_i$ wendet sich$_i$ scharf/schräg.   ('Paul$_i$ turns REFL$_i$ sharply/diagonally.')

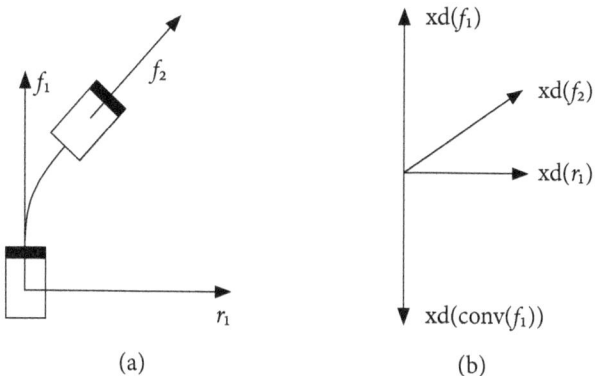

FIG. 9.10 Example for *schräg nach rechts abbiegen* with a corresponding direction diagram.

Schräg specifies that the orientation has changed slightly and *scharf* marks that the final orientation is (almost) opposite to the initial orientation. However, the exact angle is irrelevant. These adverbs give coarse information about the angle between the distinguished axis of the bearer of change before and after the change of orientation.[13] For *scharf*, the resulting front axis has to be included in the backward range between the former right axis and left axis and, thus, be more similar to *wenden*. For *schräg.* the resulting front axis should be closer to the former front axis (Fig. 9.10). Thus we can make the geometric constraints derived by adding these adverbs explicit, as in (L8) and (L9). Please note that this description is not our proposal for the semantic structure of these lexemes. It rather specifies what should be derived for these contexts from a general analysis of these terms.

(L8) scharf'$(r, srs) \Rightarrow r \in$ rd(right-axis(srs), back-axis(srs), left-axis(srs))
(L9) schräg'$(r, srs) \Rightarrow r \in$ rd(right-axis(srs), front-axis(srs), left-axis(srs))

Based on (L6) and (L8), we can characterize the change of orientation for verb phrases such as *scharf nach rechts abbiegen* ('turn sharply to the right'). This phrase says that in the resulting stage the front axis of the bearer of change is in the range delimited by the former right axis and the former back axis. (In this case, we do not need an additional axis selecting the range, since the two limiting axes are not converse.)

---

[13] Our description of the adverbs *schräg* und *scharf* considers only the combination with verbs of change of orientation. As you will see, they can be modeled similarly to *nach rechts* and *nach links* in these cases.

(L10) scharf′(r, srs) ∧ nach-rechts′(r, srs) ⇒ r ∈ rd(right-axis(srs), right-axis(srs), back-axis(srs))

The last task to solve in this chapter is to specify the ranges of direction on the geometrical level in a form such that (L10) is indeed a consequence of (L6) and (L8).

2.5. Geometric Specification of Ranges of Directions

Ranges of direction are comparable to intervals on the scale of numbers. Correspondingly, we first introduce a ternary relation of **direction betweenness** for half-lines (dbtw). The direction of r lies between that of r_1 and that of r_2 iff r is equidirected to r_1 or r_2, or there are $r_1′$ and $r_2′$ that are equidirected to r_1 and r_2, respectively, such that $r_1′$ starts at the same point as r, and $r_2′$ starts at some point on $r_1′$ different from the starting point of r, and $r_2′$, and r share a point.

(D11) $\text{dbtw}(r_1, r, r_2) \Leftrightarrow_{\text{def}} r \uparrow\uparrow r_2 \lor r \uparrow\uparrow r_1 \lor \exists r_1′, r_2′: r_2′ \uparrow\uparrow r_2 \land r_1′ \uparrow\uparrow r_1 \land \text{start}(r_1′) = \text{start}(r) \land \text{start}(r_2′) \neq \text{start}(r) \land \text{start}(r_2′) \iota r_1′ \land \exists P: P \iota r_2′ \land P \iota r$

Figure 9.11 exemplifies this definition: the half-lines $r_1′$ and $r_2′$ are drawn with dotted lines. Note that, in terms of angles, $\text{dbtw}(r_1, r, r_2)$ is valid iff the 'interior angle' between r_1 and r_2 in the direction diagram contains r. If r_1 and r_2 are converse, then every half-line r lies between them. The relation dbtw is an ordering relation with respect to a fixed first argument, i.e. it is antisymmetric and transitive.

We can now state that the half-lines in the plane are completely ordered by direction betweenness: given any two half-lines and their converse half-lines, then every half-line lies between two of these half-lines (B1).

(B1) $\forall r_1, r_2, r: (\text{dbtw}(r_1, r, r_2) \lor \text{dbtw}(\text{conv}(r_1), r, r_2) \lor \text{dbtw}(r_1, r, \text{conv}(r_2)) \lor \text{dbtw}(\text{conv}(r_1), r, \text{conv}(r_2)))$

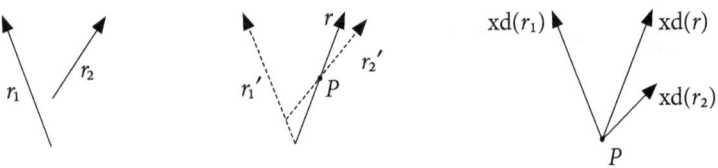

FIG. 9.11 An illustration of dbtw(r_1, r, r_2) with a corresponding direction diagram.

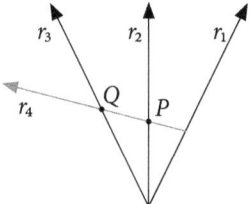

FIG. 9.12 Correspondence between dbtw(r_1, r_2, r_3) and start(r_4) $<_{r_4}$ P and P $<_{r_4}$ Q.

The ordering relations dbtw and $<$ correspond. Figure 9.12 illustrates this: in this case, dbtw(r_1, r_2, r_3) holds and any half-line r_4 that starts on r_1 and intersects the half-lines r_2 and r_3 orders the intersection points P and Q in the same way as the relation dbtw the half-lines (i.e. start(r_4) $<_{r_4}$ P and P $<_{r_4}$ Q).

Since half-lines give us exact directions, we can now use dbtw to define ranges of directions. Similar to intervals on a scale of numbers, ranges of directions can include their limiting directions. However, in the context of this article we only need **open ranges of direction**, which do not include the limiting directions.

A **range of directions** is given by three half-lines r_1, r_2, r_3 that bear the relation dbtw and where r_3 is not both converse to r_1 and parallel to r_2. The half-line r is in the range of directions given by r_1, r_2, r_3 iff r is not equidirected to r_1 or r_3 and lies between r_1 and r_2 or between r_2 and r_3.

(D12) $r \in rd(r_1, r_2, r_3) \Leftrightarrow_{def} dbtw(r_1, r_2, r_3) \wedge \neg(r_3 \uparrow\uparrow conv(r_1) \wedge r_3 \parallel r_2) \wedge \neg(r \uparrow\uparrow r_1) \wedge \neg(r \uparrow\uparrow r_3) \wedge [dbtw(r_1, r, r_2) \vee dbtw(r_2, r, r_3)]$

This specification fits the discussion of the spatial adverbial phrases given above, since it provides the basis to derive the following theorem, which presents the formal background of (L10). The theorem says that if half-line r is in the range of directions delimited by half-line r_1 and its converse and selected by half-line r_2, and r is also in the range of directions delimited by r_2 and its converse and selected by r_1, then r is between r_2 and r_1.

(T) $\forall r_1, r_2, r: r \in rd(r_1, r_2, conv(r_1)) \wedge r \in rd(r_2, r_1, conv(r_2)) \rightarrow \neg(r \parallel r_1) \wedge \neg(r \parallel r_2) \wedge dbtw(r_2, r, r_1)$

In the context of (L9) this means that if $r \in$ rd(front-axis(srs), right-axis(srs), back-axis(srs)) (nach-rechts'(r, srs)) and $r \in$ rd(right-axis(srs), back-axis(srs), left-axis(srs)) (scharf'(r, srs)), r's direction is neither equal to that of the right-axis(srs) nor to that of the back-axis(srs), but lies between them. Since we know that the right-axis(*srs*) and the back-axis(*srs*) cannot be parallel, we can apply (D12) to infer $r \in$ rd(right-axis(*srs*), right-axis(*srs*), back-axis(*srs*)).

Conclusion

The German verbs of change of orientation can be combined with various groups of adverbial phrases that specify directions. The pattern of restrictions that can be observed differs from the patterns that are characteristic for verbs of motion and verbs of orientation. The specific patterns can be ascribed to different concepts of direction encoded in the verbs. Different patterns of restrictions that we observed within the group of verbs of change of orientation can be ascribed to whether the verbs specify change of location in addition to change of orientation. However, we found that the verbs of change of orientation mark a division within the group of goal-directed adverbial phrases. This division suggests that paths and axes (**bounded** and **unbounded directions**) have to be distinguished for linguistic reasons independent of the distinction between direction of motion and object orientation.

Verbs of orientation and verbs of change of orientation access and specify spatial reference systems, which are also relevant for the interpretation of projective terms. Spatial reference systems provide a collection of axes originating at one location. Using the geometry of half-lines, we have elaborated a geometric model of axes and relations between directions. The geometric model has been applied by specifying the spatial concepts encoded in some verbs and adverbial phrases.

The geometry of half-lines is formulated in an axiomatic framework to model space and spatial relations. This framework does not use coordinates or other metrical concepts. This shows that basic relations between directions (such as parallelity, equidirection, converseness, range of directions) can be formalized without reference to either lengths, distances, or angles.

10

How Finnish Postpositions See the Axis System

URPO NIKANNE

Abstract

This chapter discusses the axis system lexicalized in the semantics of Finnish postpositions. First, it is shown that there are two kinds of postposition in Finnish meaning 'in front of' and 'behind': these postpositions indicate that both the Figure and the reference object are moving or are neutral with respect to motion. It is also shown that the Finnish prepositions meaning 'in front of' and 'behind' do not only refer to horizontal spatial relations but any one-dimensional relation. Those postpositions that mean 'beside' refer to two-dimensional relations. However, the postpositions that mean 'above' and 'below' can only refer to vertical relations. The dimensional system and the orientation of the axes (horizontal, vertical, side-to-side) are dependent on each other and should be described as parts of the same system, at least when it comes to the semantics of Finnish postpositions.

Introduction

Jackendoff and Landau (1992) show that all the spatial prepositions of English can be described by using a rather limited set of features. The system is as set out in Table 10.1 (the table is taken from Jackendoff and Landau, 1992: 119).

There are three lexical or morphological categories in Finnish that correspond to the category preposition in English: pre- and postpositions as well as locative cases. (For a discussion on Finnish locative cases and postpositions, see Nikanne, 1993.) However, the words focused on in this chapter are postpositions.

The article is divided into two parts that are related to each other. The first part of the article discusses a semantic property of the Finnish spatial postpositions that makes them different from the prepositions of English as

TABLE 10.1 *Features of spatial relations by Jackendoff and Landau (1992: 119, table 6.2)*

Reference object geometry
    Volumes, surfaces, and lines: *in, on, near, at, inside*
    Single axis
        Vertical: *on top of*
        Horizontal: *in front of, in back of, beside, along, across*
    Quantity: *between, among, amidst*
Figure object geometry
    Single axis: *along, across, around*
    Distributed figure (medium or aggregate): *all over, throughout, all along, all around, all, across*
Relation of region to figure object
    Relative distance
        Interior: *in, inside, throughout; out of*
        Contact: *on, all over; off of*
        Proximal: *near, all around; far*
    Direction
        Vertical: *over, above, under, below, beneath*
        Horizontal
             Side-to-side: *beside, by, alongside, next to*
             Front-to-back: *in front of, ahead of, in back of, beyond*
    Choice of axis system
        Inherent: *on (the) top of, in front of, ahead of, behind*
        Contextual: *on top of, in front of, behind, beyond*
    Visibility and occlusion: *on top of, underneath*
Paths (trajectories)
    Earth-oriented: *up, down, east, west, north, south*
    Figure object and axis-oriented: *forward, ahead, backward, sideways, left, right*
    Operators on regions
        Via: *through* (= via inside), *along* (= via along)
        To: *to, into* (= to in), *onto* (= to on)
        Toward: *toward*
        From: *from, from under, from inside*
        Away from: *away, away from*

described by Jackendoff and Landau (1992). Namely, in Finnish, some postpositions of PLACE meaning 'behind' or 'in-front-of' can only be used when referring to two moving objects. A corresponding group of prepositions cannot be found in English, and thus the feature is missing from the features in the inventory by Jackendoff and Landau. I will argue that the difference between the semantics is made possible because of grammatical differences between these two languages. Thus, to a certain extent, the grammar of the particular language may determine what kinds of meaning are (typically) expressed.

The latter part of the chapter suggests that the use of prepositions can be used as a basis for a theory of a hierarchical axis system. Using Finnish data, I will modify the model of axial system developed by van der Zee (1996). In the discussion, the features 'vertical', 'horizontal', 'side-to-side', and 'front-to-back' are put in a new light. It appears that at least when it comes to the semantics of Finnish postpositions, 'horizontal' is a derived and not a primitive feature—or to put it in another way: the horizontal axis (as it is expressed in the Finnish language) receives its orientation from the vertical axis.

1. Stationary versus Moving Objects

In conceptual semantics, PLACE is understood as a spatial (or abstract) region that is related to some object, the **reference object**. And where there is a PLACE (or PATH) there also is a **Theme**, the **Figure** that is located in the PLACE (or moving along the PATH) (see Nikanne 1990, 2000a). The terminology is illustrated in Fig. 10.1, in which the sentence *John is standing next to the table* is used as an example sentence.

The semantics of the English preposition system sees PLACES as stationary entities. However, there is no a priori reason to assume that both the **Figure** and the **Ground** (the **Theme** and PLACE) could not be in motion. It appears that PLACES as stationary entities is only a language specific property of the English preposition system (and this probably holds for other Germanic languages as well). It is possible to express a relation between two objects such that one is behind the other. One may also want to indicate that the objects are

FIG. 10.1 Theme, PLACE, and reference object. Illustration of the sentence *John is standing next to the table*.

moving such that their mutual relation stays the same. In English one needs to use a motion verb: *John is following behind Bill*. In Finnish this may be done also by just using particular **postpositions**. Consider the data in (1) and (2):

(1) a. Buick on Volvon edellä.
    Buick is Volvo+GEN in-front-of
    'The Buick is driving such that it stays in front of the Volvo.'
  b. Buick on Volvon edessä.
    Buick is Volvo+GEN in-front-of
    'The Buick is in front of the Volvo.'

(2) a. Buick on Volvon perässä/jäljessä.
    Buick is Volvo+GEN behind
    'The Buick is following behind the Volvo.'
  b. Buick on Volvon takana.
    Buick is Volvo+GEN behind
    'The Buick is behind the Volvo.'

In Finnish, for instance the postpositions *edellä* 'in front of', *perässä* 'behind', and *jäljessä* 'behind' (1a, 2a) indicate that both the Buick and the Volvo are moving. The postpositions *edessä* 'in front of' and *takana* 'behind' (1b, 2b) are neutral with respect to the stationary or moving status of the objects.

The example in (3) shows that the postpositions *edellä*, *perässä*, and *jäljessä* cannot be used to indicate a relation between two stationary objects:

(3) Maija istuu Villen takana/*perässä/*jäljessä//edessä/*edellä.
    Maija sit+3SG Ville+GEN behind//in-front-of
    'Maija is sitting behind//in front of Ville.'

However, the postpositions *edessä* 'in front of' and *takana* 'behind' can be used with moving objects:

(4) Maija kulkee Villen takana/perässä/jäljessä//edessä/edellä.
    Maija go+3SG Ville+GEN behind//in-front-of
    'Maija is following behind//going such that she stays in front of Ville.'

As the translations of the Finnish examples suggest, the English grammar requires that a motion verb is used in order to indicate motion.

In English, it is not enough to use a motion verb in order to express a PLACE in motion. Only a verb such as *follow*, which indicates that the relationship of the object is not changing, can be used with the prepositions *behind* and *in front of*. An ordinary motion verb (*go, move, drive, walk, run*, etc.) indicates that the situation changes: *The Buick drove in front of the Volvo* means roughly the same as *The Buick overtook the Volvo*, in other words, the goal of the movement is the place in front of the Volvo. We will return to this shortly in section 2.

2. The Finnish Prepositions, Postpositions, and Locative Cases

In order to understand why Finnish postpositions are able to make a distinction between stationary and moving objects, we must take a brief look at the grammatical properties of Finnish pre- and postpositions and locative cases. Syntactically, they all belong to the same syntactic category, (P)reposition.

According to Nikanne (1993), Finnish has two kinds of Ps: phonologically empty and overt ones.

2.1. Overt Ps

There are two kinds of overt p-positions in Finnish. Most of the Ps are postpositions,[1] with a projection PP of the following form:

[NP+genitive [P]]

For instance the postpositions in examples (1)–(4) belong to this group and it is the most common type of P, and thus the type of PPs (prepositional—or postpositional—phrases) mostly discussed in this chapter.

The other type of P has its complement in the partitive case. Most of these Ps can appear either before or after their complement phrase. The form of the PP-projection of these Ps is:

[[P] NP+partitive] or [NP+partitive [P]]

For example: Koira juoksee kotia kohti / kohti kotia.
 dog run+3SG home+PAR toward / toward home+PAR
 'The/A dog is running toward home.'

---

[1] The NP selected by the postposition is in the genitive case and it triggers possessive agreement in the head P. Both the case marking and the agreement seem to indicate that the complement NP is in the specifier position in the PP. It should thus be noted that the term postposition is only referring to the surface order of the words in the p-positional phrase (PP). Finnish postpositions of this type often used to be nouns that have grammaticalized as postpositions. For instance *jäljessä* 'behind' (see examples 2–4) is the inessive form of the noun *jälki* meaning 'trace' or 'track'. Thus, the expression *someone*+GEN *jäljessä* has meant concretely 'to be on someone's track'. As the genitive is the case for NP-specifiers in Finnish, it is natural that the syntactic construction NP+GEN N is easily reanalyzed as NP+GEN P when the head noun has a suitable meaning. (Cf. in English PPs such as *in front of something* where the noun *front* has lost its 'nounness' and has become a part of the preposition.)

Phonologically empty Ps govern the so-called semantic cases (i.e. locative cases and cases such as instructive 'with', comitative 'with', and abessive 'without'). The locative cases correspond to the most basic prepositions in Germanic languages; prepositions such as 'in', 'on', 'at', 'from', 'to', and 'into'. I am aware that the 'group of the most basic prepositions' is not a well-defined concept, but it seems to be somehow easy to grasp.[2]

For us, the most important property of the locative cases is that they can be combined with lexical postpositions. For instance the postpositions used in examples (1) through (4) are actually inflected in locative cases. This is typical in Finnish: Ps have a paradigm with two to six locative cases. Table 10.2 gives the case paradigms of the postpositions *ete-* and *taka-*. Note that the stem *ete-* cannot occur without the case suffix, and the stem *taka-* can only occur as the first part of a compound word (e.g. *takaovi* 'rear door', *takapää* 'rear end', etc.).

TABLE 10.2 *The case paradigm of the postpositions* ete- *and* taka-

ede+ssä
in-front-of+INESSIVE (The meaning of the inessive case is 'in'.)

ede+stä
in-front-of+ELATIVE (The meaning of the elative case is 'from' or 'from inside'.)

ete+en
in-front-of+ILLATIVE (The meaning of the illative case is 'into'.)

ede+llä
in-front-of+ADESSIVE (The meaning of the adessive case is 'at' or 'on' depending on the object; see Nikanne, 1990; 19–23; Groundstroem, 1988.)

ede+ltä
in-front-of+ABLATIVE (The meaning of the ablative case is 'from a place that can be indicated with the adessive case'.)

ede+lle
in-front-of+ALLATIVE (The meaning of the allative case is 'to a place that can be indicated with the adessive case'.)

taka+na
behind+ESSIVE

taka+a
behind+PARTITIVE

taa+kse
behind+TRANSLATIVE

[2] For instance, this is how I explain the meaning of the Finnish locative cases to my Norwegian-speaking students in the first year Finnish class when they are learning the locative cases. They have no difficulty getting the idea. But, I don't have any good theoretical explanation why 'on' and 'in' are more basic than, for instance, 'under' or 'beside'.

Historically the partitive has been a locative case indicating separation ('from') and the essive has been a spatial locative case meaning 'at' and the translative a spatial meaning 'to'.[3]

The possibility of inflecting p-positions in locative cases that also indicate basic spatial relations makes it possible to have finer semantic distinctions expressed by PPs. In English (and in Norwegian, for example), spatial prepositions such as 'behind' and 'in front of' cannot be governed by spatial prepositions referring to the goal-direction (to, toward), and therefore the meaning must rely more on the governing verb.

Compare the Finnish examples in (5) and (6). (The translative case in (6b) has historically been used as a spatial locative case meaning 'to' but in modern Finnish indicates a more abstract transition; cf. the note on the essive case above.)

(5) a. Buick ajoi Volvon eteen/edelle.
Buick drove Volvo+GEN in-front-of+ILL/in-front-of+ALL
'The Buick drove in front of the Volvo (i.e. overtook it).'
b. Buick ajoi Volvon edessä/edellä.
Buick drove Volvo+GEN in-front-of+INE/in-front-of+ADE
'The Buick drove such that it stayed in front of the Volvo'

(6) a. Buick ajoi Volvon taakse/perään.
Buick drove Volvo+GEN behind+TRA/in-front-of+ILL
'The Buick drove behind the Volvo' (i.e. came close to it).'
b. Buick ajoi Volvon takana/perässä.
Buick drove Volvo+GEN behind+ESS/in-front-of+INE
'The Buick was following behind the Volvo.'

There is a general rule that a movement verb indicates a change of location. And, because in Finnish it is not possible to use the locative case that indicates

---

[3] In modern Finnish, the essive is still used but it indicates a more abstract relation: normally having some property, role, or state of mind, e.g.

(i) Martti Ahtisaari oli Suomen presidenttinä vain yhden kauden.
Martti Ahtisaari was Finland+GEN president+ESSIVE only one+ACC term+ACC
'Martti Ahtisaari was the president of Finland for only one term.'
(ii) Minä muutin Norjaan vanhana.
I moved+1SG Norway+ILLATIVE old+ESSIVE
'I was old when I moved to Norway.'

The translative case has a similar meaning but it means 'to the state that can be indicated with the essive case':

(iii) Ahtisaari valittiin Suomen presidentiksi vuonna 1994.
Ahtisaari elect+PASS Finland+GEN president+TRANSLATIVE year+ESSIVE 1994
'Ahtisaari was elected as the president of Finland in (year) 1994.'
(iv) Olen tullut vanhaksi.
be+1SG come+PAST-PARTICIPLE old+TRANSLATIVE
'I have become old.'

location and not path, we are able to end up with a meaning that both of the objects in question are moving (changing location) but their mutual spatial relation stays the same and the reference object (the one governed by the preposition) is not the goal of the Theme.

In English, only one spatial relation can be expressed by a preposition, and it is necessary to fix the meaning of the construction with a movement verb plus a preposition indicating location. It seems that the fixed meaning is that the place expressed by the preposition *in front of* or *behind* is the goal of the movement—and the meaning thus corresponds to that of the Finnish examples (5a) and (6a), with the locative cases indicating goal.

(7) The Buick drove in front of the Volvo (i.e. overtook it).
(8) The Buick drove behind the Volvo (i.e. came close to it).

The rule in English could be formulated in a very sketchy way as:[4]

(9) $[_{VP} \ldots V_i \ldots PP_j \ldots]$
    MOTION$_i$ TO PLACE$_j$

The notation should be understood as follows. A PP referring to a PLACE (co-indexed with the index j) will be governed by a PATH-function TO—which does not have a lexical counterpart in the syntactic structure—if the PP is selected by a motion verb in the syntactic structure (MOTION and V co-indexed). As this is a fixed—and in principle unpredictable—linking between form and meaning, we could say that (9) is a '**construction**' in the sense of Goldberg (1995) or Fillmore and Kay (1997).

3. Three Groups of Postpositions

Penttilä (1957) has a rather exhaustive list of Finnish pre- and postpositions (among other things). Taking his listing as my corpus, I could divide the spatial

---

[4] A more sophisticated formalization within the formalism developed in Nikanne (1990, 1995, 2000b) is:

$[_{VP} \ldots V_i \ldots [_{PP} \ldots P_j \ldots [NP_k] \ldots] \ldots]$

$\quad\quad [\ldots] \quad\quad\quad\quad [\ldots]_k$
$\quad\quad\ \ |\quad\quad\quad\quad\quad\ \ |$
$\ldots f2_i \rightarrow f1 \rightarrow f1_j$
$\quad\quad\ \ |\quad\quad\ |\quad\quad\ |$
$\quad\quad\ Dir\quad Dir\quad\ \ \ldots$
$\quad\quad\quad\ \searrow\swarrow$
$\quad\quad\quad\ \ $goal

('Dir' stands for directionality, '...' stands for unspecified features, an arrow indicates selection and a line indicates the association of features and other elements of the structure. For more information of this formalism, see Nikanne (1990, 1995, and 2000b).)

postpositions indicating a spatial PLACE into three groups with respect to their ability to indicate a spatial relation of two moving objects:

TABLE 10.3 *Finnish postpositions that indicate a spatial* PLACE

Neutral:
   alla 'below', 'under', 'underneath'
   alapuolella 'below'
   edessä 'in-front-of'
   keskellä 'in-the-middle-of'
   keskessä 'in-the-center-of'
   kohdalla 'in-the-same-location-as'
   luona 'by'
   lähellä 'near'
   paikkeilla 'approximately-in-the-same-location-as'
   puolella 'on X's side'
   päällä 'on', 'over'
   rinnalla 'beside'
   takana 'behind'
   tienoilla 'approximately-in-the-same-location-as'
   tykönä 'with', 'by', 'close to'
   ulkopuolella 'outside-of'
   ulottuvilla 'in-the-reach'
   vierellä 'beside'
   vieressä 'beside'
   välillä 'between'
   yllä 'above', 'over'
   ympärillä 'around'
Implicate movement:
   edellä 'in-front-of', 'ahead', 'before'
   hännillä 'in-the-last-position-of (a moving group)' (lit.: 'on the tails')
   jäljessä 'behind' (lit.: 'in the trace/track/footprint' cf. n. 1)
   kannoilla 'very close behind' (lit.: 'on the heels')
   perässä 'behind'
Implicate stative NP-complement:
   edustalla 'in-the-front-of'
   huipulla 'on-the-top-of'
   juurella 'at-the-foot-of'
   juuressa 'at-the-foot-of'
   liepeillä 'close-to'
   suunnalla 'in-the-approximate-location-of'
   äärellä 'by'
   ääressä 'by'

Most of the postpositions are neutral with respect to a motion of the Theme and the reference object and some indicate that the objects move but not with respect to each other. There is also a group of postpositions that indicates that the complement NP corresponds to a stationary object. This is intuitively not very surprising: it is very common in everyday life that one has to relate an object (a Theme) to another object whose location is stable. For instance a typical example of the postposition *edustalla* would be *kaupan edustalla* (shop+GEN in-the-front-of) and an example of *liepeillä* would be for instance *kaupungin liepeillä* (city+GEN close-to), etc.

4. Vertical Movement

One interesting thing is that all the Finnish postpositions that indicate motion of the objects correspond to the English prepositions *in-front-of* and *behind*. All the **vertical axis prepositions** *alla* 'under', *alapuolella* 'below', *yllä* 'over', *yläpuolella* 'above', *päällä* 'on/over', are neutral with respect to the moving or stationary status of the objects. Consider the examples in (10):

(10) a. Haukka on varpusen yläpuolella/yllä/päällä//alapuolella/alla.
       hawk is sparrow+GEN above/over/on//below/under
       'The hawk is above/over//below/under the sparrow.'
      b. Haukka lentää varpusen yläpuolella/yllä/päällä//alapuolella/alla.
       hawk fly+3SG sparrow+GEN above/over/on//below/under
       'The hawk is flying above/over//below/under the sparrow.'

The examples in (10a) are neutral with respect to movement. In (10b), the sentence has a motion verb instead of a simple copula. The meaning is that the hawk and the sparrow are flying horizontally but the flying course of the hawk is above that of the sparrow.

It does not seem natural to express the mutual relation of two vertically moving objects using the vertical postpositions. Imagine two rockets, rocket A and rocket B flying on the same course as illustrated in Fig. 10.2. If we want to express the situation in Finnish, we have to find the right postpositions:

(11) a. Raketti A on/kulkee raketin B perässä/jäljessä/takana
       rocket A is/go+3SG rocket+GEN B behind
      b. ??Raketti A on/kulkee raketin B alapuolella/alla
       rocket A is/go+3SG rocket+GEN B below

(12) a. Raketti B on/kulkee raketin A edellä/edessä
       rocket B is/go+3SG rocket+GEN A in-front-of

b. ??Raketti B on/kulkee raketin A yläpuolella/yllä/päällä
rocket B is/go+3SG rocket+GEN A above/over

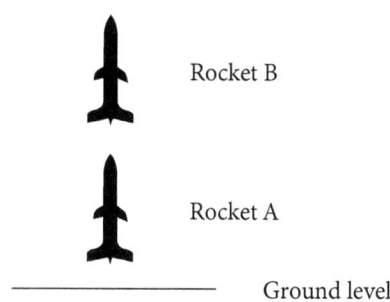

FIG. 10.2 Two rockets in motion.

It is striking that the vertical postpositions are very unnatural if not ungrammatical in this case.[5] The only natural way to use postpositions in this case is to use those that correspond to the English prepositions *in front of* and *behind*. The obvious question is, of course, what is it in these postpositions that makes them special?[6]

5. Postpositions Meaning 'Behind' and 'In Front Of' Expressing Horizontal Movement

The first intuition in relation to the postpositions meaning 'behind' and 'in front of' is that they basically express a horizontal relationship. This is also how the English prepositions *behind* and *in front of* are analyzed by Jackendoff and Landau (1993) (see Table 10.1 above). Sentences (13) and (14) show how this horizontal relationship is expressed by these postpositions:

(13)   Helikopteri A lensi helikopterin B yläpuolella ja helikopteri C lensi
       B:n edellä. D lensi A:n jäljessä.
       chopper A flew+3SG chopper+GEN B above and chopper C flew
       B+GEN in-front-of. D flew A+GEN behind
       'Chopper A was flying above chopper B and chopper C was flying in front of B. D was flying behind A.'

---

[5] Some speakers of Finnish find these examples acceptable if they imagine a three dimensional space as a setting of the situation, i.e. that not only the vertical path is important but that something interesting is going on around the two rockets.
[6] Michael Carter (pers. comm.) has confirmed to me that English works here in the same way as Finnish. Only *behind* and *in front of* are allowed in such a situation as that in (11). Actually, he was surprised and thought that it a peculiarity of English.

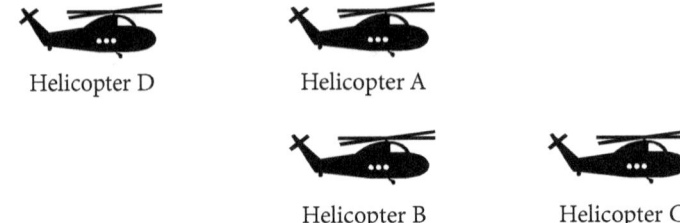

Fig. 10.3 Flying helicopters.

(14) a. Katsomossa Ville istuu Maijan yläpuolella, Pekka hänen edessään, Erkki takana ja Anna alapuolella.
auditorium+INE Ville sit+3SG Maija+GEN above, Pekka she+GEN in-front-of, Erkki behind and Anna below
'In the auditorium, Ville is sitting above Maija, Pekka in front of her, Erkki behind and Anna below.'

However, the data in (11) and (12) show that the postpositions referring to English *in front of* and *behind* do not always refer to the horizontal plane.

Note that in examples (14b) and (14c) the orientation of the front–back axis need not be horizontal:

(14) b. Kärpänen lentää mehiläisen vieressä/vierellä/rinnalla
fly fly+3SG bee+GEN beside
'The fly is flying beside the bee.'

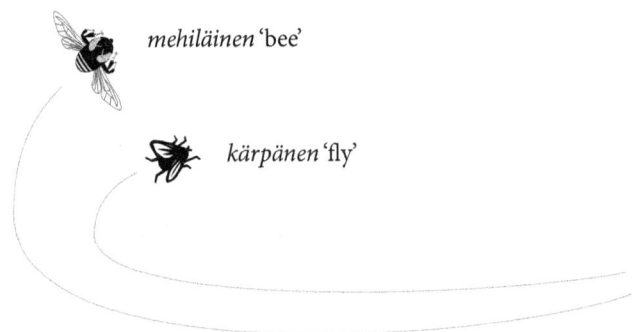

FIG. 10.4 Kärpänen lentää mehiläisen vieressä/vierellä/rinnalla.
fly fly+3SG bee+GEN beside
'The fly is flying beside the bee.'

(a)

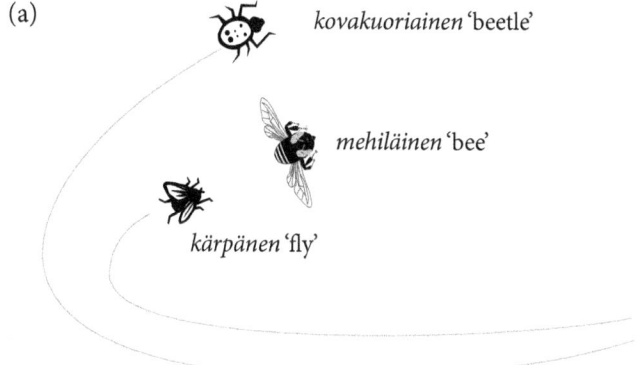

FIG. 10.5 (a) Mehiläinen on/lentää kärpäsen edellä ja kovakuoriainen sen vieressä vierellä.
bee is/fly+3SG fly+GEN in-front-of and beetle it+GEN beside
'The bee is flying in front of the fly and the beetle (is flying) beside it.'

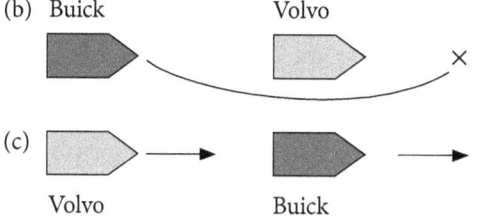

(b) Buick ajoi Volvon eteen/edelle
Buick drove Volvo+GEN in-front-of+ILL/in-front-of+ALL
(c) Buick ajoi Volvon edellä/edessä
Buick drove Volvo+GEN in-front-of+ADE/in-front-of+INE

(14) c. Mehiläinen on/lentää kärpäsen edellä ja kovakuoriainen sen vieressä/vierellä/rinnalla/sivulla.
bee is/fly+3SG fly+GEN in-front-of and beetle it+GEN beside
'The bee is/is flying in front of the fly and the beetle is/is flying beside it.'

Of course, the sentences in (11a) and (12a) can also express a non-vertical path of movement as long as rocket A is following rocket B on the same path of movement (just as the fly is following the bee in example (14b). The examples in Fig. 10.5 (b) and (c) illustrate how case can modify a motion verb, such that the Figure either moves to a location in front of the reference object or is in front of the reference object while both Figure and reference object are moving.

It is possible to use the expression in (14b) whenever the bee and the beetle are flying in the same direction and one is not ahead of the other—and, if they are not on the same vertical axis. The side-to-side axis (postpositions *vieressä/ vierellä/rinnalla/sivulla*) is the axis that is perpendicular (90 degrees) to the front–back axis. (There is also a further restriction for the use of the postpositions *vierellä* and *vieressä*: the inherent side-to-side axes of the two participants (the fly and the bee) must be more or less aligned. This condition is not as strong with the postposition *sivulla*.)

In the following section, we will take a closer look at the axis system expressed by the Finnish postpositions.

6. Axial System of PLACE-Postpositions in Finnish: Integrating Dimension and Orientation

It has become clear from previous sections that the postpositions meaning 'behind' and 'in-front-of' do not necessarily express horizontal spatial relations, although they may do so. I assume that the fundamental semantics of these postpositions is to express a relationship between two objects on a **one-dimensional directed axis**. The horizontal meaning can be derived from the axial system. By the term '**axis**' I mean a directed line or half-line that can divide a space into different dimensions. I am operating only with the conventional three-dimensional space with three axes—one vertical and two horizontal ones.

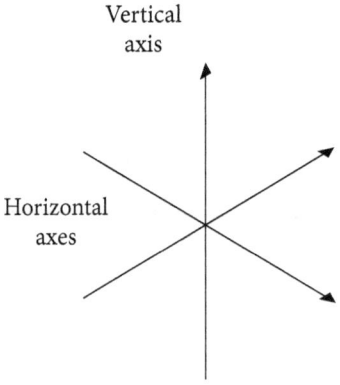

FIG. 10.6 (a) Axial system: one vertical and two horizontal axes.

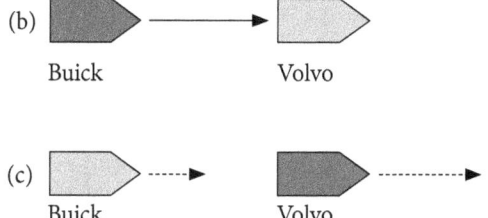

FIG. 10.6 (b) Buick ajoi Volvon taakse/perään.
Buick drove Volvo+GEN behind+TRA/in-front-of+ILL
(c) Buick ajoi Volvon takana/perässä.
Buick drove Volvo+GEN behind+ESS/behind+INE

Van der Zee (1996) has studied the **axis system** as it appears in the use of certain Dutch directional prepositions and prepositional phrases. Van der Zee's theory of reference frame is a very good starting point for us; it is illustrated in (15). ('f1' is standing for function of zone 1, i.e. a function that can derive a PLACE or PATH out of a reference object.)

(15) The reference frame in van der Zee (1996)

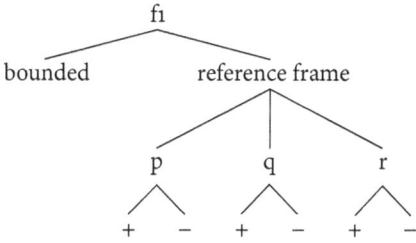

p = front–back axis
q = left–right axis
r = bottom–top axis
(p and q can also stand for east–west, north–south, etc. axes)

The model in (15) assumes that all the axes (p, q, and r) are at the same hierarchical level. However, the behavior of the Finnish postpositions suggests that **the reference frame** is more hierarchical than in (15), and we therefore need to modify van der Zee's model. Note that the reference frame in (15) cannot explain the special status of the vertical axis or the front–back axis because all axes are hierarchically equal. In addition, the hierarchy in (15) tacitly assumes a three-dimensional space as all the three dimensional axes are immediately dominated by the same node ('reference frame').

I suggest that the reference frame is a feature hierarchy illustrated in (16). (The Finnish words *ylä-*, *ala-*, *etu-*, and *taka-* are used because the hierarchy is based on the Finnish language.)[7]

(16) Feature hierarchy for Finnish spatial postpositions

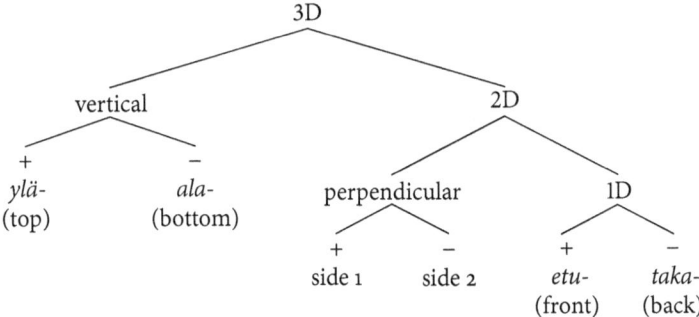

The idea is to tie **dimensionality** together with the horizontal versus vertical status of the axes. According to the hierarchy in (16), the vertical axis implicates three-dimensionality. The **vertical axis** is special because it is the only axis whose orientation is defined independently of the other axes' orientations. Note that also Bryant, Tversky, and Lanca (2000) show that the subjects in their experiments were particularly fast at determining object locations in the above–below dimension. The linguistic data from Finnish also point out a special status for the vertical axis: the orientation of the vertical axis is always consistent. The **horizontal orientation** of the other two axes can be derived from the vertical axis.

In Bryant *et al.*'s research subjects were also rather fast in the front–back dimension: they were quickest at reporting remembered object positions at the top–bottom dimension, followed by the front–back dimension, and were least fast with respect to the side–side dimension. Also in (16), the front–back axis has a special status relative to the other axis in the horizontal plane: its inter-

---

[7] It should be noted that (16) is based on linguistic data from Finnish. I have not run any other experiments besides using my own native speaker's intuition about the interpretation of the examples such as those given in this chapter. I have also asked other native speakers' judgements. If one would like to run a more sophisticated psycholinguistic experiment, it could be done for instance on a computer screen with 3-D graphics. The subjects should tell in one sentence what is going on in the moving or stationary picture on the screen. One could have different tests: one where the subjects could use only the verb *olla* 'be' (in order to get the effects of the choice of preposition) and another one in which the subjects could use any verbs they like (in order to see the effect of the choice of verb). In addition, one could ask the subjects to draw situations described in sentences such as those in this chapter. However, experimenting in this way I leave for future research.

pretation is independent of the other axis. The **side-to-side axis** can only be interpreted, however, in relation to either the **front–back axis** alone or in relation to the front–back axis and the top–bottom axis together.

The front–back axis is neutral with respect to horizontality or verticality when the vertical axis is not involved. The axis under 2D is perpendicular to the other axes. As it is in the middle, it follows that all three axes are perpendicular to each other when they are present. In principle, the axes under 3D and 1D could be identical but this is ruled out by a separate condition: the axes must not have an identical orientation. If the vertical axis is present, the front–back axis is interpreted to be horizontal. This is because it, too, must be perpendicular to the vertical axis.

Now it should be clear why only the postpositions meaning 'in-front-of' and 'behind' are used in examples (1) through (4). The front–back axis is the directed 1-D axis. The path the objects are moving along in examples (1) through (4) and (12) and (13) is exactly what we are looking for: a one-dimensional directed axis. Because no other dimensions are necessary to assume as parts of the expressed situation illustrated in (10) through (12) with the two rockets, it is the most natural choice to use the postpositions that express 'in-front-of' and 'behind'.

Notice that the special status of the vertical axis is not a logical necessity. If it was the case that the horizontal axis was the special axis for the postposition semantics, the postpositions meaning 'above' and 'below' would be used to refer to spatial relations on a one-dimensional directed axis. Instead of 'in front of', we would say 'above' (or 'below') even if the motion was horizontal—for instance one animal chasing another (*Hey, the dog is right below the cat!* meaning 'Hey, the dog is right behind the cat!') (cf. the data in 11–12). This is not the case in Finnish (or English), however.

It is not surprising that the vertical axis is the one whose orientation is fixed and that the orientations of the other axes depend on the vertical axis. In addition to the facts discussed in Bryant, Tversky, and Lanca (2000; see the discussion above), in ordinary life, the vertical orientation is less dependent on varying deictic factors: gravity is basic, and people and other objects only very seldom appear in an upside-down position. The horizontal axes are less fixed. We could also say this in another way: The vertical axes of people and other objects (the 'what-system vertical axes') most of the time have the same orientation and this is also the orientation of the vertical axis based on gravity (the 'where-system vertical axis').

(16) is based on a **feature hierarchy**—not much different from that of modern phonology—and is designed having axes and not **vectors** (see Zwarts 2000) in mind. As far as I can see, the feature system could as well be applied

to the vector formalism because it can deal with direction, dimension, and orientation. The suggested feature system is only making a claim about the relationship between dimensions and within the semantics of Finnish PPs. If there is a principled difference between the axis and vector formalisms, it goes beyond the problems discussed in this chapter.

Conclusion

I have shown that in Finnish some postpositions that express a stationary spatial relation between two objects (i.e. PLACE) implicate that the objects are moving—but not with respect to each other. In this respect, Finnish differs from English and the Finnish data help us to understand better the possibilities of the lexical semantics of p-positions.[8] All these postpositions correspond to the English prepositions *in front of* (*edellä*) or *behind* (*perässä, jäljessä, kannoilla, kintereillä*). This fact is understandable, assuming that the postpositions meaning 'in front of' and 'behind' express a one-dimensional relation and are not specified to be horizontal. The front–back axis is horizontal only if the vertical axis is assumed to play a role in the situation. This fact can be derived if the whole axial system is something likethat suggested in (16).

[8] Interestingly enough, this fact of Finnish postpositions has not been pointed out in the Finnish grammatical literature earlier even though all the native speakers I have discussed this phenomenon with (and there are many) agree with me about the interpretation of the meaning of the prepositions *takana* vs. *perässä/jäljessä* and *edessä* vs. *edellä*. Possibly we Finnish grammarians see our own language through the Germanic/Romance grammatical tradition, without noticing it.

11

Directions from Shape: How Spatial Features Determine Reference Axis Categorization

EMILE VAN DER ZEE and RIK ESHUIS

Abstract

This chapter presents three experiments on Dutch directional nouns, like *voorkant* 'front', and prepositions, like *voor* 'in front of'. The first two experiments focus on the intrinsic meaning of these terms in relation to the horizontal plane. The experiments show that the following spatial features of a reference object determine reference axis categorization: (a) axis length, (b) contour expansion, and (c) curvature of the main plane of symmetry. On the basis of the first two experiments, as well as insights by Clark (1973), Tversky (1996) and Landau and Jackendoff (1993), the Spatial Feature Categorization (SFC) model is formulated. This model generates predictions on reference axis categorization derived from the spatial features of a reference object for the purpose of intrinsic directional reference in both the horizontal and the vertical plane. A third experiment tests predictions by the SFC model in relation to Dutch directional nouns by using a new set of reference objects, a new experimental task, and by considering both the horizontal and the vertical plane. This experiment supports predictions by the SFC model. It is argued that a system of categorized reference axes is necessary to represent local direction for the purpose of intrinsic directional reference. The categorization of the axes in such a system can be explained on the basis of the spatial features of a reference object.

We would like to thank the Department of Psychonomics at the University of Utrecht in the Netherlands for providing the means for carrying out the third experiment.

Introduction

It is generally assumed that the meaning of directional expressions, like *behind the bed* or *north of Lincoln*, is based on a system of categorized reference axes (see e.g. Carlson-Radvansky and Irwin, 1993; Carlson, Regier, and Covey, Ch. 6; Eschenbach, Habel, and Lessmoellmann, 1997; Jackendoff, 1996; Landau and Jackendoff, 1993; Levinson, 1996; Logan and Sadler, 1996; Slack and van der Zee, Ch. 1; Tversky, 1996; van der Zee, 1996). Depending on the linguistic context a reference axis is categorized, for example, as a front–back axis or a north–south axis, and the location of a **Figure** is subsequently described or understood in relation to such an axis (van der Zee, Rypkema, and Busser, 1996). The intrinsic meaning of a directional expression derives from the categorization of reference axes based on the intrinsic properties of a **reference object** or **Referent**. Intrinsic properties of Referents are such properties as the way in which a Referent is categorized (e.g. what the object affords us to do with it, what its **function** is, or what its role is in a particular context; see Carlson, 2000; Coventry, 1998, Ch. 13; Gibson, 1950; Herskovits, 1986; Vandeloise, 1991), its **force dynamic properties** (e.g. its **center of gravity** and its **streamline**; see Regier, 1996; Talmy, 1988, 2000), its **part structure** (see Harris and Strommen, 1974), and its **orientation** or **movement** in relation to other objects (see Nikanne, Ch. 11). This chapter considers how Dutch directional nouns like *voorkant* ('front') and prepositions like *voor* ('in front of') can be used to refer to the location of a Figure by taking into account the spatial properties of a Referent (e.g. its curvature and its part structure).

The next section describes an experiment exploring which spatial properties of a reference object determine the intrinsic use of Dutch directional prepositions in the horizontal plane.

1. Spatial Features Determine how Dutch Directional Prepositions Refer to Regions in the Horizontal Plane

The set of three dimensional reference objects used in experiments 1 and 2 is depicted in Fig. 11.1. These six reference objects differ from each other (a) in whether the cross-section expands along the **main axis** or not, i.e. **expansion** or **no expansion** (the main axis or **generative axis** is the **longest axis** and/or the **symmetry axis**, which means that this axis can be curved), and (b) in the amount of curvature of the **main plane of symmetry**, i.e. **not curved** (0°), **slightly curved** (45°), or **highly curved** (180°) (the **main plane of symmetry**

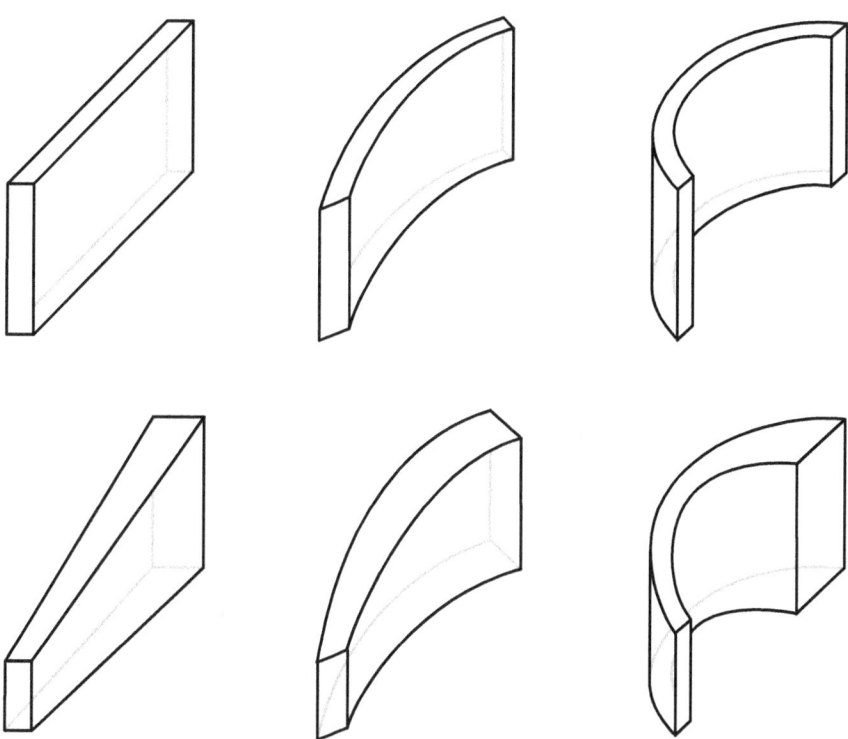

FIG. 11.1 The reference objects used in experiments 1 and 2.

is defined here as the largest plane around which a reference object is mirror symmetric). These **spatial features** are paraphrases of the defining parameters of what Biederman (1987) calls **geons**: a finite set of basic object parts that are assumed to play a pivotal role in visual object recognition (Dickinson *et al.*, 1997). Although Biederman's **Recognition By Components (RBC) theory** only assumes one axis representation in each geon, namely the main axis, two additional axis representations are assumed here that are orthogonal to each other and to the main axis. All axes intersect each other in the geometric mid-point of a geon. The three axes of this local axial system are in principle able to represent three different directions: the **left–right, top–bottom**, and **front–back direction**. (This chapter does not consider the categorization of **axis parts**, like the front part of a front–back axis. As reviewed by Landau (Ch. 2), there is much evidence to suggest that the cognitive mechanism categorizing axes is different from the cognitive mechanism categorizing axial parts.)

The first two experiments explore whether there is a systematic influence of the above-mentioned spatial features on **axis categorization** for the purpose of **intrinsic directional reference** in the **horizontal plane**. Even though these spatial features follow from RBC theory, the experiments described here do not test RBC theory, nor does the outcome of the experiment depend on RBC theory. Taking object parts as Referents in the experiments described here offers the advantage, however, of studying a possible influence of a well-defined set of spatial features on axis categorization without any interference by the part structure of the Referents. Any reference to geons instead of reference objects in this chapter is therefore meant to be pragmatic, and does not necessarily indicate a commitment to RBC theory.

The **distribution** in space of intrinsic directional regions or object sides is explained here in terms of a different axial system than the **categorization** of such regions or sides. As just indicated, the categorization of intrinsic directional regions and object sides is based on the intrinsic axial system of the object involved. However, it is assumed here that the distribution in space of such regions and object sides is based on the intrinsic axes of a **closest fitting cuboid** or a **bounding box** around each reference object. This idea is illustrated in Fig. 11.2.

Using the axes of a bounding box to represent the spatial distribution of three orthogonal dimensions corresponds to the intuition that the left–right, top–down, and front–back dimensions are ideally orthogonal to each other in

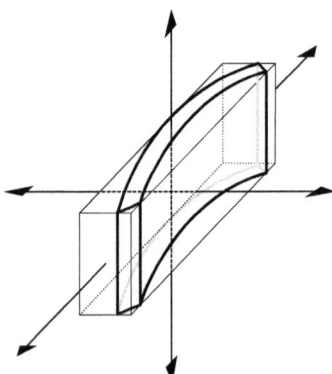

FIG. 11.2 It is assumed here that the distribution of the intrinsic regions and object sides that correspond to Dutch directional prepositions and nouns is based on the position and orientation of the main axis and two orthogonal axes of the closest fitting cuboid or bounding box around a reference object.

space. The idea of using a bounding box model to explain such a distribution is controversial, however. For example, Regier and Carlson (forthcoming) and Carlson, Regier, and Covey (Ch. 6) argue that a bounding box model for generating *above* regions does not compare favorably with predictions by their own **Attentional Vector Sum (AVS) model**. For pragmatic reasons, however, the bounding box model is used here as an approximation to the true distribution of directional regions. By distinguishing between the distribution of directional regions and object sides (on the basis of intrinsic bounding box axes) and the categorization of such regions (on the basis of categorized object axes), it is possible to focus on the contribution of spatial features to **reference axis categorization**, while considering the **distribution of directional regions and objects sides** as a separate research issue.

In the first experiment ninety-six Dutch speaking participants were asked to put two dots on a black circle surrounding each of the six reference objects (at least 23cm away from the outer edge of each geon), either *voor* ('in front of') and *achter* ('behind') the reference object, or *links van* ('to the left of') and *rechts van* ('to the right of') the reference object. So, for each reference object a participant either indicated the representation of a front–back axis or a left–right axis. Participants indicated the referential status of each dot by putting a 'v' (for *voor*), and an 'a' (for *achter*), or an 'l' (for *links van*) and an 'r' (for *rechts van*) next to it. While facing the left side of a toy car participants were reminded, before the start of the experiment, that there is a difference between referring to regions *links van de auto* ('to the left of the car') and *voor de auto* ('in front of the car') 'from their own perspective' or 'from the perspective of the car'. None of the participants had problems understanding this distinction. The participants were asked to put the dots on the circle 'in the same way as when they took the perspective of the toy car'. In order to enhance the chance of using the intrinsic meaning of the above-mentioned directional prepositions the participants walked twice around the table on which the objects were displayed (thus constantly changing their own perspective in relation to the objects), and put the dots and markings on the circle on their second round. A white blanket around the experimental set-up prevented the participants from using any directional cues from the experiment room. In order to determine—after the experiment—whether directional cues from the experiment room had been used, the stimuli were displayed in what might be called a standard orientation in half of the trials, and were displayed at a 45° orientation from the standard orientation in the other half of the trials. After the experiment the participants completed a questionnaire in which they were asked whether they had associated the reference objects with any known objects, and whether such associations had helped them to determine their responses.

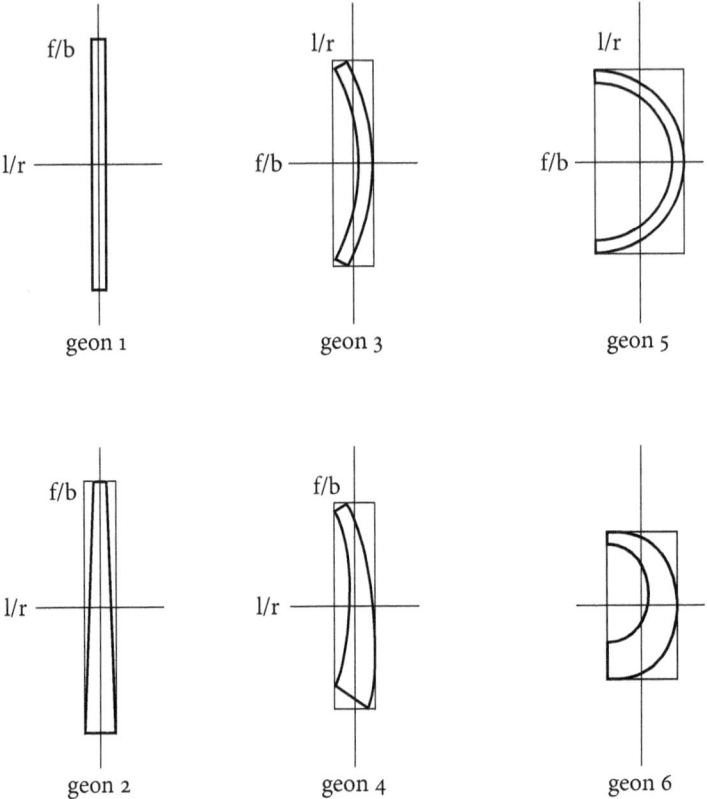

FIG. 11.3 Cuboid axis categorizations for intrinsic reference to regions around reference objects in the horizontal plane based on the use of Dutch directional prepositions. F/b indicates 'front–back axis' and l/r 'left–right axis'. There is no difference in the categorization of the cuboid axes for geon 6. Stimuli not to scale.

In 80.1 per cent of the cases the dots were placed in regions around the reference objects that are defined by projecting bounding box sides away from the Referent, along one of the bounding box axes, whereas the probability of doing so was on the average 25 per cent (see van der Zee (1996) for further details).[1] By considering the labeling of the dots that fall in these regions, certain **cuboid axis categorizations** can be deduced (see Fig. 11.3).

---

[1] The a priori chance of scoring according to the hypothesis is between only 1.5 and 2.5% for each geon if one takes into account that there must always be two scores on opposite sides of each geon that are categorized in opposite ways.

The results indicate that curvature of the main plane of symmetry, **Referent axis length** and expansion have an influence on cuboid axis categorization. For geon 1, which has a non-curved main plane of symmetry, and no expansion along the main axis, participants consider the cuboid axis contiguous with the longest Referent axis as the front–back axis, and the cuboid axis contiguous with the shortest Referent axis in the horizontal plane as the left–right axis. For geons 3 and 5, however, which do have a curved main plane of symmetry, and also lack expansion along the main axis, the cuboid axis contiguous with the Referent axis orthogonal to the curved plane of symmetry is categorized as the front–back axis. The influence of contour expansion along the main axis becomes clear by comparing the cuboid axis categorizations for geons 1, 3, and 5 on the one hand, and geons 2, 4, and 6 on the other hand. Depending on the curvature of the geons involved, the categorizations of the cuboid axes differ for geons with an expanding contour and a non-expanding contour (see geons 3 versus 4, and 5 versus 6). This result also indicates an interaction between curvature and expansion. Such an interaction is confirmed by considering only geons with an expanding contour. Comparing geon 6 on the one hand, with geons 2 and 4 on the other hand, shows that whereas the two cuboid axes of highly curved geon 6 are equally likely to be categorized as a front–back axis and a left–right axis, these axes receive different kinds of categorizations for non-curved geon 2 and slightly curved geon 4.

The presentation of the results on cuboid axis categorization assumes that cuboid axis categorization is determined by **geon axis categorization**, which in turn depends on spatial features like geon axis length, orthogonality of a geon axis to a curved plane of symmetry, and contour expansion along a geon axis. In order for geon axes to have such an influence, a **transfer of geon-axis-to-cuboid-axis categorization** must be assumed. The properties of such a transfer are theory-dependent, however. Although a bounding box model would assume a geon-axis-to-cuboid-axis-categorization transfer, a model like AVS would have to assume a **geon-axis-categorization-to-vector-categorization transfer** in order to explain how vectors representing local direction are categorized. Because of the theory dependence of a transfer mechanism, cuboid axis categorization is assumed to depend here on geon axis categorization without further addressing how such a transfer might work.

The possibility that the results presented in Fig. 11.3 might have been caused by environmental factors can be excluded. The orientation in which the reference objects were presented to the participants did not have an influence on the way the participants represented the axes in a reference object, or on the way in which they categorized these axes.

Of the subjects, 28 per cent indicated that associations with known objects had helped them to determine their responses. Objects that were mentioned most, however, namely 'wall' (18 per cent of the total number of associations) and 'building' (11 per cent of the total number of associations), could not have had an influence on where the dots were placed. For example, doors in buildings are not consistently found in the smaller side of an elongated building, which could have given participants cues for the placement of the dots around the smaller side of geon 1. Neither can the categorization of the dots have been influenced by the associations that were mentioned. If, for instance, the 'wall' association had played a role, one would expect a reverse labeling of the axes on, e.g., geons 1 and 2 (where the regions bordering on the largest side would then have been categorized as IN FRONT OF and BEHIND, since Dutch native speakers can refer to the regions bordering on the longest side of a wall with 'in front of the wall' or 'behind the wall', but not to the regions bordering on the shorter sides of a wall). This means that—although our questionnaire does not make it possible to rule out the influence of features other than spatial ones—it does seem to make it possible to exclude the influence of conscious associations the participants may have had with the stimuli that were used.

It can be concluded that experiment 1 shows a systematic labeling of dots, corresponding to the location of what Dutch native speakers consider to be 'in front of', 'behind', 'to the left of', and 'to the right of' a reference object, based on spatial features like relative axis length, relative curvature of the main plane of symmetry, and contour expansion along a reference object's main axis.

Let us now consider how Dutch native speakers use directional nouns in relation to object sides.

2. Spatial Features Determine How Dutch Directional Nouns Refer to Object Sides in the Horizontal Plane

In experiment 2 the same reference objects are used as in experiment 1. In this experiment participants put stickers either on the *voorkant* ('front') and *achterkant* ('back'), or on the *linkerkant* ('left side') and *rechterkant* ('right side') of the reference objects. The arrows in Fig. 11.4 indicate expected sticker placements if participants put their stickers as close as possible to the **intrinsic reference object axes**, the position of which is defined in the introduction.

The categorization of the intrinsic Referent axes is assumed to be identical to that in experiment 1. For example, in the case of geon 5 one sticker may be used to indicate the FRONT of the object, and one sticker to point out the BACK of the object, at opposite midpoints of the longest object sides. These sticker placements are called **real side predictions**.

Directions from Shape 217

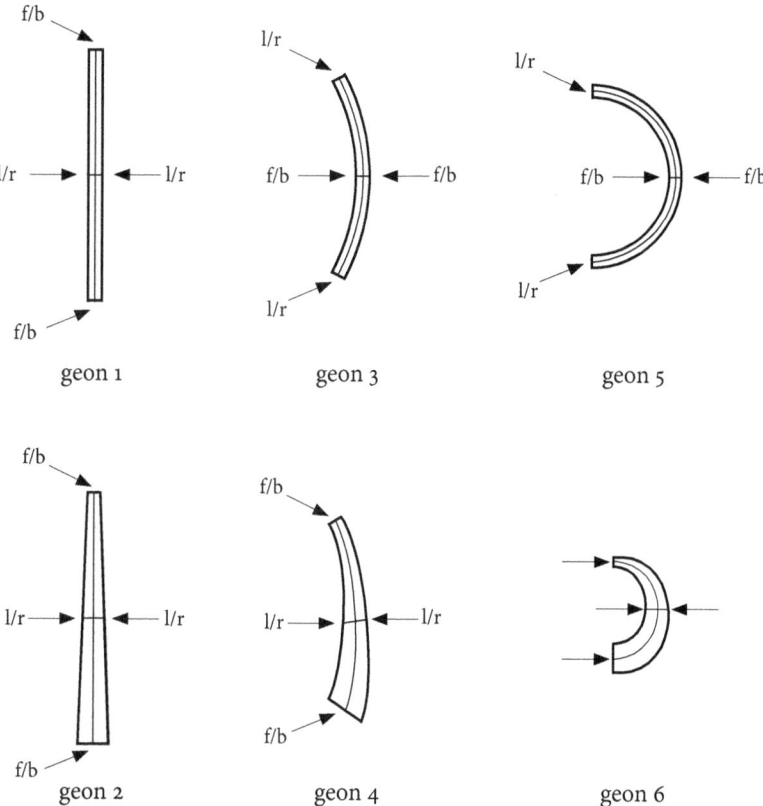

FIG. 11.4 Predictions of sticker placements corresponding to Dutch directional nouns, assuming that the sticker placements are based on the intrinsic axes of the reference objects that are categorized as in experiment 1 (real side predictions). Stimuli not to scale.

It is possible, of course, that participants do not use the intrinsic reference object axes to represent the distribution of directional object side locations. Let us consider what predictions follow from the bounding box model, given that the bounding box model gives an acceptable approximation of the true distribution of directional object side locations. If participants should want to place their stickers on the virtual sides of a bounding box, this would not be possible, since participants have only the real sides available. However, it could be assumed that if participants want to place their stickers on a virtual side they will choose that location on a real side that is as close as possible to a virtual side. For example, participants who want to indicate the FRONT and the BACK of geon 5, and want to put their stickers on the virtual front and back of the bounding box would now need three stickers: for example, two stickers to indicate the front, and one sticker to indicate the back, because there are two locations at which the virtual front touches a real object side, but only one location at which the virtual back touches a real object side. Such predictions are called **virtual side predictions**.

Fig. 11.5 shows the virtual side predictions for geons. It is assumed here that the sides of the bounding boxes—and thus the locations indicated in Fig. 11.5—inherit their categorization from the cuboid axes that intersect the bounding box sides.

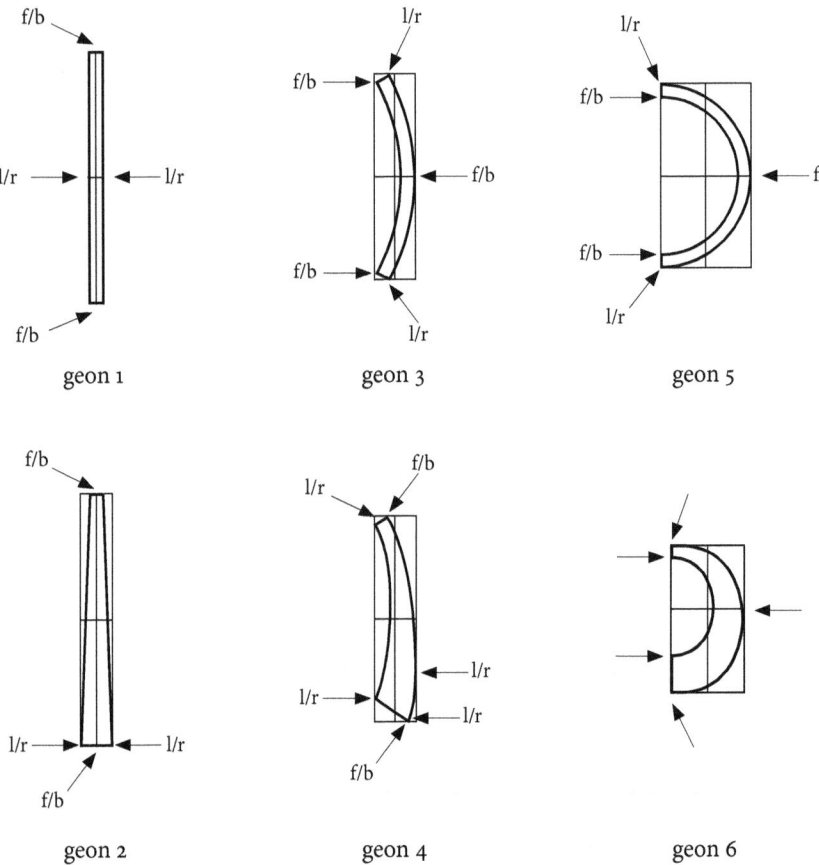

FIG. 11.5 Predictions of sticker placements corresponding to Dutch directional nouns, assuming that the sticker placements are as close as possible to the side of a closest fitting cuboid. The categorization of the sticker positions is based on the categorization of the cuboid sides, which inherit their categorization from the cuboid axes as in experiment 1 (virtual side predictions). Stimuli not to scale.

It follows from Figs. 11.4 and 11.5 that it is possible to make a distinction between real side predictions and virtual side predictions for some of the geons on the basis of the number of stickers used and where the stickers are placed (e.g. compare the distinct real side and virtual side predictions for geon 5 in Figs. 11.4 and 11.5).

In experiment 2 all participants were presented with eight stickers: two referring to the *voorkant* ('front'), two referring to the *achterkant* ('back'), two referring to the *linkerkant* ('left side'), and two referring to the *rechterkant* ('right side'). Also in this case the objects were presented to the participants on a round table, with a minimum of environmental cues. All objects were presented in either a standard orientation or in a 45° orientation from the standard. A toy car was used to explain the experimental task to the participants. None of the participants had problems indicating the intrinsic *voorkant* ('front') and *linkerkant* ('left side') of the car. Participants could take as many stickers as they wanted to indicate two opposing directional object side locations. They placed their stickers during their second tour around the table. And, as was also the case in relation to experiment 1, they completed a questionnaire in which they were asked whether any associations between these objects and known objects had helped them to determine their responses.

The results show that real side predictions and virtual side predictions together account for 82 per cent of all the sticker placements, whereas the a priori probability of scoring according to both predictions is 43 per cent (see Eshuis 1995 for more details).[2] Of all scores that are according to both hypotheses, it is not possible to determine in 28 per cent of the cases whether the scores are real side scores or virtual side scores, because both hypotheses can account for these scores. However, of all scores that can be accounted for in terms of either one hypothesis (72 per cent of all correct scores), 82 per cent of the scores are real side scores, and 18 per cent are virtual side scores. These results show that both real side scores and virtual side scores adequately predict the placement and categorization of stickers referring to directional object sides. This means that the categorization of the dot placements involved depends on the spatial features of the reference object in exactly the same way as found in experiment 1 (even though the way in which the distribution of the dot placements was derived for real side and virtual side predictions differs).

Only 19 per cent of the subjects indicated that associations with known objects did have an influence on how they placed the dots. The most men-

---

[2] The a priori chance of scoring according to both hypotheses is only 6.27% if one takes into account that the scores must always be on opposite sides of each geon, and must always be categorized in opposite ways.

tioned associations were 'wall' (22 per cent of the total number of reported associations) 'fish' (20 per cent of the total number of reported associations), and 'building' (9 per cent of the total number of reported associations). For similar reasons as mentioned in experiment 1 it also does not seem possible that the associations with 'wall' or 'building' could have played a role in determining the participants' responses in experiment 2. If 'fish' associations had played a role the only possible candidates are some of the scores in relation to geons 2 and 4. However, even if all scores on these geons are considered in relation to the total number of 'fish' associations the maximum impact is only 4 per cent of the total scores. Given the total number of real side and virtual side scores this seems negligible.

Also in this experiment the orientation of the reference objects in relation to the environment did not affect scoring, thereby excluding any influence of directional cues from the environment.

Let us consider next how the systematic influence of spatial features on the use of Dutch directional prepositions and nouns can be described best.

3. Spatial Features Determine How Dutch Directional Nouns Refer to Object Sides in the Horizontal and the Vertical Plane

Experiments 1 and 2 show that the following spatial features have an influence on intrinsic directional reference: (a) the length of an intrinsic reference object axis in relation to the length of other axes, (b) whether an axis of a reference object is orthogonal to the main curved plane of symmetry or not, and (c) whether there is expansion of a geon's contour along its main axis or not. It is possible to formulate a model that explains the systematic influence of these spatial features on reference axis categorization. This section describes such a model, as well as an experiment testing predictions by this model in relation to a set of reference objects, another experimental task, and an additional dimension in which intrinsic directional reference can take place.

Van der Zee and Eshuis (1997) describe the so-called **Spatial Feature Categorization (SFC) model**. This model takes the spatial features of a reference object as its input and delivers categorized axes—for the purpose of **directional reference**—as its output. Landau and Jackendoff (1993) assume that intrinsic reference object axes are **directionally marked** for the purpose of intrinsic directional reference. The first component of the SFC model elaborates on this idea by assuming that intrinsic reference object axes are directionally marked on the basis of the spatial features of a reference object. The second component of the SFC model specifies in what way directional axis marking corresponds

TABLE 11.1 *Correspondence between directional axis marking and lexical concepts describing reference axis categorization for intrinsic directional reference by Dutch native speakers*

Most directionally marked	BOVEN–ONDER ('TOP–DOWN') axis
Intermediately directionally marked	VOOR–ACHTER ('FRONT–BACK') axis
Least directionally marked	LINKS–RECHTS ('LEFT–RIGHT') axis

to **axis categorization**. Research by e.g. Clark (1973) and Tversky (1996) can be interpreted as pointing out that the correspondence between directional axis marking and axis categorization is systematic in English. We assume the same systematicity for Dutch (see Table 11.1).

The details of the two components of the SFC model may differ from language to language. It seems reasonable to assume that different languages may have different ways of directionally marking the local axes of a reference object. For example, speakers of Tzeltal may refer to the location of a Figure in relation to reference object parts that are idealized as body parts, as in 'the plate is at the ear side or belly side of the table', corresponding to what English speakers may refer to as the 'corner' and the region 'below' a table (Levinson, 1992*b*). Apart from a different distribution of relevant object parts for directional reference compared to English or Dutch, it may be assumed that speakers of Tzeltal employ a different system of directional axis marking than native English or Dutch speakers, since in the latter case the relative importance of 'ear parts' does not have to be established in relation to a 'belly part'.

The second component of the SFC model is language-specific in the sense that different languages employ different lexical concepts that correspond to the different levels of directional axis marking. For example, what is categorized as the FRONT–BACK axis in English is categorized as the VOOR–ACHTER axis in Dutch. However, despite the fact that the details of the two components of the SFC model may differ across languages, the SFC model assumes that directional axis marking and a systematic correspondence between directional axis marking and axis categorization are universal components of directional linguistic reference.

The SFC model makes it possible to derive some interesting new predictions. The empirical findings have so far been confined to the horizontal plane. However, by treating directional marking as a process that cuts across dimensions the SFC model also makes predictions about axis categorization in the **vertical plane**. This becomes clear if we reformulate the influence of two spatial features in terms of the SFC model: (a) increasing axis length leads to increased directional marking, and (b) an axis that is orthogonal to a main curved plane of symmetry receives more directional marking than other axes in the same intrinsic axial system. This reformulation entails that if only axis length is considered, a longer axis is directionally marked more than a shorter axis, and is therefore more likely to be categorized as a TOP–DOWN axis than a shorter axis. The same idea applies to an axis that is orthogonal to a curved plane of symmetry. Also such an axis is directionally marked more than other intrinsic reference object axes, and is therefore also more likely to be categorized as a TOP–DOWN axis than another axis in the same axial system. In the experiment described below these predictions are tested.

The SFC model allows for the derivation of another set of predictions as well. This follows from the fact that—in principle—the SFC model also allows other spatial features to have an influence on directional axis marking and thus on reference axis categorization. Harris and Strommen (1974) discuss another likely candidate. These authors show that adding an object part, such as a small circle, to a simple geometric form, like a square, induces participants to assign a FRONT and a BACK to such objects (where the circle feature is considered to be at the FRONT of an object, and where the opposite side is considered to be the BACK of the object). Harris and Strommen's experiments only focus on the horizontal plane, however. By reformulating their findings in terms of the SFC model it can be stated that an axis that intersects a salient object part is more directionally marked than an axis that does not intersect such a part, and that the former axis is therefore more likely to be categorized as a TOP–DOWN axis than another axis. By reformulating the findings of Harris and Strommen in terms of the SFC model predictions can thus be derived about the possible role that object parts may play in intrinsic directional reference in both the horizontal and vertical dimension. Also these predictions are tested in experiment 3, below.

The stimuli shown in Fig. 11.6 were designed to test predictions by the SFC model. All stimuli that are used are hollow. The three cylinders at the bottom left of Fig. 11.6 are full cylinders that are hollow, those at the bottom right are

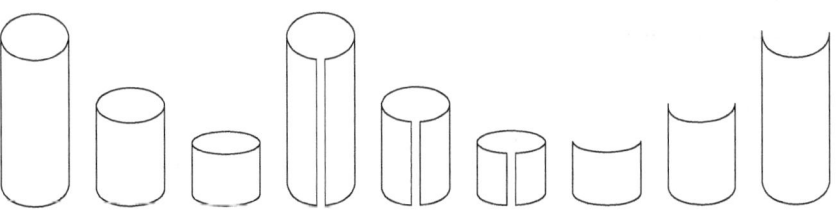

FIG. 11.6 The stimuli that were used in experiment 3. Not to scale.

half cylinders that are hollow, and those grouped together in the middle are hollow cylinders into which a groove has been cut, so that a seven-eighths cylinder is the result. All cylinder types come in three different lengths.

The properties of the cylinders make it possible to investigate predictions by the SFC model. The effect of axis length can be determined across cylinder types. In addition to this, the three full cylinders make it possible to establish an effect of axis length in isolation from any of the above-discussed other spatial features. This is the case whether one interprets the full hollow cylinders in terms of their positive space (their 'pipe-likeness') or in terms of their **negative space** (their 'hollowness'). If one considers the full hollow cylinders in terms of their negative space there is one unique symmetry axis that coincides with the midpoint of the hollow cross-section. This symmetry axis is longer than the ambiguous axis of the negative cross-section. However, if one considers the full hollow cylinders in terms of their positive space, there are many axes parallel to the main negative cylinder axis that are longer than the axes representing the width of the pipe. In other words, no matter whether one interprets the full hollow cylinders in terms of their negative or positive space, it is possible to establish the contribution of a set of longer axes in relation to a set of shorter axes.

The three half-cylinders in the bottom right corner of Fig. 11.6 make it possible to determine whether the axis orthogonal to the main curved plane of symmetry is more likely to be categorized in terms of a TOP–DOWN axis than a FRONT–BACK axis or a LEFT–RIGHT axis. The three seven-eighths cylinders at the top of Fig. 11.6 make it possible to determine whether a groove is more likely to be categorized in relation to the TOP–DOWN axis than in relation to the FRONT–BACK axis or LEFT–RIGHT axis. In fact, if an effect in relation to the TOP–DOWN axis is found, this can be due either to considering the groove as an object part consisting of negative space, or to the presence of a curved plane of symmetry in the positive object part. The axis that is orthogonal to the main curved plane of symmetry namely coincides with the axis intersecting the **negative object part**. In other words, if an effect is found, this can be interpreted either as a confirmation for negative object parts playing a role in intrinsic directional reference, or of a curved plane of symmetry playing such a role.

Seventy-two Dutch participants received the three cylinders, three half-cylinders and three seven-eighths cylinders on a supporting frame. The frame ensured that all participants first encountered the stimuli at a 45° angle in relation to gravity. The participants were asked to take each object from the supporting frame and hold it in front of them, with the object's *voorkant* ('front') toward their own front, and the object's *bovenkant* ('top') pointing upwards—thereby also indicating the left–right dimension of the object. The participants could rotate each object in as many ways as necessary to determine the required properties of the object. Each participant rehearsed the task with the same toy car that was used in the first two experiments. No one had problems understanding this task. After the experiment the participants also completed a questionnaire in which they were asked whether they had associated these objects with any known objects, and whether such associations had helped them to determine their responses.

The results show that there is an overall influence of axis length on the way the participants orient the stimuli (χ^2: 10.30, df = 4, $p \leq .036$), as well as an influence of shape (χ^2: 32.12, df = 4, $p < .0301$).[3] These results confirm some of the expectations based on the SFC model: relative axis length does have an influence on how participants orient the objects in three-dimensional space. Also different spatial features, represented by the three different object types, have an impact on the way participants orient the objects in three-dimensional space (about which more below). A more refined analysis of the relative orientation for each object type in three-dimensional space reveals an interesting pattern. Placement of the main axis of the whole cylinder differs significantly in relation to the participants' top–down axis, their front–back axis, and their left–right axis (χ^2: 28.584, df: 2, $p < .001$). Participants most often align the longest axis with their own top–down axis, less often with their own front–back axis, and least often with their own left–right axis. This pattern confirms the expected effect of axis length. The longer an axis is, the more directionally marked it is, and the more likely it is to be categorized in terms of TOP–DOWN compared to FRONT–BACK or LEFT–RIGHT. There is also a significant difference in the placement of the groove of the seven eighths cylinder in relation to the body axes of the participants (χ^2: 28.584, df: 2, $p < .001$). Participants most often align the axis intersecting the negative object part with their own top–down axis, less often with their own front–back axis, and least often with their own left–right axis. This pattern confirms the expected effect of object part saliency, or the effect of the main curved plane of symmetry. This outcome thus supports either the role of negative object parts, or of main curved planes of symmetry in relation to intrinsic directional reference. Finally, there is a significant difference in the way in which the curved plane of symmetry of the half-cylinder is placed in relation to each of the three body axes (χ^2: 65.334, df: 2, $p < .001$). Participants most often align the axis that intersects the curved plane of symmetry with their own top–down axis, less often with their own front–back axis, and least often with their own left–right axis. Also this pattern confirms predictions by the SFC model: An axis that intersects a

---

[3] Due to the distribution of the data it is not possible reliably to determine any interactions between variables, such as between length and object curvature.

curved plane of symmetry is more directionally marked than any other axis.

Did associations with known objects help any of the participants? This does not seem to be the case. Out of all reported associations 37 per cent referred to objects whose main axis is usually oriented along the horizontal (e.g. 'drainpipe', 'tunnel', and 'bridge'). Nineteen per cent referred to objects whose main axis is usually aligned with the gravity vertical (e.g. 'vase', 'can', and 'wall'). And 44 per cent of all associations referred to objects whose main axis may be oriented along any axis (e.g. 'cylinder', 'pipe', and 'case'). These figures indicate that associations with known objects cannot have caused the pattern of results found. If associations had been instrumental in the scoring of our participants we should expect more alignments along the front–back and left–right axes (horizontal axes) than along the top–down axis. This is not the case.

Conclusions

Directional nouns and prepositions can be said to refer to a distribution of categorized regions in three-dimensional space. For Dutch native speakers these are regions like *voor x* ('in front of x') and *de voorkant van x* ('x's front'). It is possible to explain the distribution of these categorized regions in terms of **axes** (Landau, Ch. 2), **bounded half lines** (Schmidtke, Tschander, Eschenbach, and Habel, Ch. 9), **vectors** (Carlson, Regier, and Covey, Ch. 6; O'Keefe, 1996, Ch. 4; Zwarts, 1998, Ch. 3) or **topological distinctions** (Gambarotto and Muller, Ch. 8). But whatever kind of representation is chosen, a set of reference axes is always necessary to represent local direction for the purpose of intrinsic directional reference. Such a system of axes categorizes the relevant regions in terms of a set of lexical concepts representing direction.

Here we have presented empirical evidence for the Spatial Feature Categorization Model. This model assigns lexical concepts to the axes of a reference frame on the basis of the spatial features of a reference object: (a) the length of a reference axis in relation to the length of other local reference axes; (b) whether a reference axis is orthogonal to an object's main curved plane of symmetry or not; (c) whether there is expansion of a reference object's contour along the main reference object axis or not; and—possibly—(d) whether an intrinsic reference object axis intersects a salient object part or not (here: a negative object part). We feel that any theory on directional reference must specify how the spatial features of a reference object contribute to directional reference. Only in this way is it possible fully to understand and appreciate the contribution of other kinds of reference object features, such as functional features, affordances, and contextual factors.

12

Memory for Locations Relative to Objects: Axes and the Categorization of Regions

RIK ESHUIS

Abstract

This chapter discusses the spatial information that directional expressions refer to. The focus is on the intrinsic object-centered use of these expressions. An idea for the correspondence between spatial representations and linguistic reference to such representations is presented for the 2-D case. Two experiments test this idea. The results suggest that axes are used to carve up space surrounding a Referent into regions. The observed regional error points to the representation of axial boundaries between half-planes. This finding contradicts the representation of axial prototypes (Hayward and Tarr, 1995) as well as the representation of axial boundaries between quadrants (Crawford, Regier, and Huttenlocher, 2000). Although vectors may define specific relations between Figure and Referent, the position relative to the Referent's axes determines the membership of a Figure's location to opposing intrinsic regional categories. The best examples of directional prepositions are spatial relations which can be defined with a vector that coincides with a half-axis.

Introduction

Spatial language may be used to communicate information concerning the location of objects. In English and in other languages such as Dutch and German one way of describing an object's location is by means of a **directional**

The author thanks Emile van der Zee and Annette von Wolff for valuable comments on earlier drafts of this chapter.

preposition such as *in front of* (Dutch: *voor*, German: *vor*). Directional prepositions specify spatial relations between objects. A sentence such as *the bike is in front of the house* expresses the location of one object (the **Figure**) relative to another object (the **Referent**). With respect to the Referent (*the house*) a direction (*to the left of*) is defined in which the Figure (*the bike*) can be found (see e.g. Talmy, 1983). The spatial relations characterized by the directional prepositions *above*, *below*, *in front of*, *behind*, *to the left of*, and *to the right of* will be called **directional relations** here. This chapter considers the representation of directional relations.

Directional prepositions do not necessarily convey unambiguous information regarding the position of a Figure. It may be unclear how the preposition in the above example has to be interpreted because the relation may have been defined from, e.g., the viewpoint of the speaker or from the viewpoint in which houses are normally encountered. This possible ambiguity is caused by the different **frames of reference** in which directional relations can be defined. See Levinson (1996) for one way to classify frames of reference. Here, reference frames will be distinguished by (1) whether directions are extracted from the Referent itself or whether directions are projected onto the Referent (**intrinsic** vs. **relative** reference respectively) and (2) whether the source of the directional information is the ego, another object, or the environment (respectively **ego-centered**, **object-centered**, and **environment-centered** reference). This chapter focuses on intrinsic object-centered frames of reference. Referring in an intrinsic object-centered manner means specifying a Figure's location with a direction that is based on inherent features of the Referent. A viewpoint-invariant direction is thus defined by some intrinsic aspect of the Referent's structure that does not depend on any other perspective. Objects with inherent properties that are suited as a Referent are objects that have *tops*, *bottoms*, *fronts*, *backs*, and/or *sides*.

Directional prepositions do not refer to precisely defined spatial relations, that is, they do not differentiate between Figure locations within a part of space adjacent to the Referent. *The bike* in the above example might be located at various distances from *the house* and closer to the left side than to the right side of the house or vice versa. A relation expressed by *in front of* can be applied to various Figures in a **region** defined relative to a Referent. Directional spatial relations are categorical in that specific Figure positions within the region can be treated equally (Landau and Jackendoff, 1993; Talmy, 1983). This does not mean that a particular preposition is used indifferently over such a region: A speaker may sometimes use a combination of directional prepositions and the applicability of a given preposition may vary over the region (Hayward and

Tarr, 1995; Logan and Sadler, 1996). To understand this aspect in more detail it is necessary to consider how such regions are constructed around objects, and how a relation between a Figure and a Referent is categorized, i.e. how it is established that the specific location a Figure occupies belongs to a particular region. Regarding intrinsic object-centered frames of reference the additional question is how object features can be used to define regions that are linked to the Referent's structure. These questions are the main topic of this chapter.

In the next section, some—partially contradicting—ideas and findings concerning the correspondence between linguistic reference to directional relations and the structure of the spatial information representing these relations are discussed. As opposed to spatially representing **axes as prototypes** (Hayward and Tarr, 1995) or **axes as boundaries** between prototypical **quadrants** (Crawford, Regier, and Huttenlocher, 2000) the approach that is promoted here for intrinsic object-centered reference is that of **axes as boundaries between half-planes**. **Half-axes** are the best examples of the half-planar regions when a directional preposition is used to refer to this spatial information structure. In section 2, the results of two experiments are presented. The experiments are designed to test spatial memory for locations of a Figure in an intrinsic object-centered frame of reference. In section 3, the results are discussed and compared with the approach advocated in section 1.

1. Spatial Language and Spatial Representation

Spatial language is often assumed to refer to or to build upon (non-linguistic) spatial representations (Landau and Jackendoff, 1993; Landau, Ch. 2). Therefore, it is also assumed that 'any aspect of space that can be expressed in language must also be present in non-linguistic spatial representations' (not necessarily the other way around; Landau and Jackendoff, 1993: 217). These assumptions entail that on the basis of, for example, visual information, sentences containing directional prepositional phrases can be uttered, and the other way around, upon hearing such sentences information can be derived which may be helpful in a visual search. The manner in which spatial information characterizing directional relations is represented may thus exert an influence on linguistic as well as non-linguistic spatial behavior. This makes it possible to investigate the connection between spatial language and spatial representation by comparing results on linguistic and non-linguistic spatial tasks. In this section, some studies that make such comparisons are considered in more detail.

1.1. Axes as Prototypes

To shed some light on the underlying principles of linguistic and non-linguistic spatial relations Hayward and Tarr (1995) report four experiments. Similar constellations of Referents and Figures were used in linguistic as well as non-linguistic spatial tasks. One of the constellations of the Figure (a circle) and the Referent (a computer monitor) that was used in all four experiments is shown in Fig. 12.1. The grid defines the places where the Figure could be located and was not visible to the participants at any time during the experiments. The Referent was always in the center (row 4, column 4). For present purposes it suffices to consider only experiments 1 and 3, in which—respectively—spatial relations were described and reproduced. In experiment 1 participants had to describe the position of the Figure relative to the Referent. In experiment 3 the task of the participants was to remember the location of the Figure with respect to the Referent and reproduce the spatial relation: A configuration of the Referent and Figure was briefly presented, would then disappear, and only the Referent would reappear. The participants had to put the Figure back in the location where it was before by using a computer mouse.

In experiment 1 Hayward and Tarr found that participants used a single directional preposition when the Figure was positioned directly vertical (*above/below* in column 4) or horizontal (*to the left of/to the right of* in row 4)

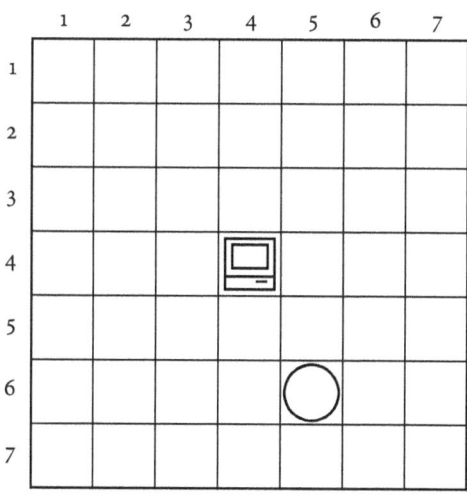

FIG. 12.1 One of the stimulus configurations used by Hayward and Tarr (1995).

relative to the Referent. When the Figure was in another position, mostly a combination of two directional prepositions was used. To avoid ceiling effects only the first preposition that was mentioned by a participant was used to analyze the relative amounts of use of the different prepositions over the locations around the Referent (see Hayward and Tarr 1995: 51, sect. 2.4). In this way they found the highest percentage use of vertical terms (*above/below*) and horizontal terms (*left/right*) in column 4 and row 4 respectively, with gradually decreasing usage with distance from the—conjectured—axes. In experiment 3 accuracy of the Figure position estimates followed a similar pattern. On the horizontal locations (row 4) the error in the vertical direction was smaller compared to locations in the other rows, on the vertical locations (column 4) the error in the horizontal direction was smaller compared to locations in the other columns.

The main rationale of Hayward and Tarr's work is that if particular spatial relations can be shown to have an effect on a non-linguistic spatial task such as the reproduction task, then this is evidence for the organization of spatial representations that is independent of the use of spatial prepositions. Because the same distribution of effects is found in a linguistic as well as a non-linguistic spatial task, Hayward and Tarr make the assumption that 'Spatial relations in non-linguistic systems and spatial predicates in language both encode spatial forms as prototypes' (p. 76) and that this correspondence is not accidental. Hayward and Tarr conjecture that spatial information represented as **axial prototypes** causes the same pattern of results on linguistic as well as non-linguistic spatial tasks.

From Hayward and Tarr's experiments it is not clear which kind of spatial information is responsible for the effects found. The observed axial system is defined in relation to several superimposed frames of reference: An intrinsic object-centered frame (based on the Referent), a relative ego-centered frame (based on the participant), and a relative environment-centered frame of reference (based on the gravitational vertical and/or the computer or the sheet of paper on which the stimuli were presented). The experiments described in section 2 are designed to investigate spatial relations in a purely intrinsic object-centered frame of reference by manipulating the orientation of the Referent, so that only the intrinsic reference frame can be used. Note further that Hayward and Tarr found that mostly two prepositions are used to describe Figure locations that are not on an axis (for similar results, see Zimmer, Speiser, and Baus, 2001). At these locations an object is, for example, not only *above* but *to the left* as well. At a location on an axis, on the other hand, an object is, for example, *above* but not *to the left of* nor *to the right of* the object. Thus what Hayward and Tarr consider to be the (prototypical) **above–below axis** may also be interpreted to function as the **boundary** between **left** and **right regions**. Not only can findings on linguistic tasks be reinterpreted as such, findings on non-linguistic spatial tasks may also suggest axial boundaries as is shown next.

1.2. Axes as Boundaries

Crawford, Regier, and Huttenlocher (2000) present findings that point to an inverse relation between spatial language and spatial representation. In their experiments 1 and 2, participants first rated the applicability of *above* with respect to Figure–Referent configurations, and then reproduced the spatial relations by putting the Figure back in its former position by using a computer mouse.[1] In addition to accuracy of the reproduction Crawford *et al.* also consider the **bias of the distributions of Figure positions**, i.e. the direction and distance of the mean of the reproduced Figure positions from the true location of the Figure. What they found is that reproduction is relatively accurate on the Referent's axes and that the less accurate reproduction estimates on the remaining test locations show a bias away from the axes and toward a position in between the axes. The results are interpreted according to the model of Huttenlocher, Hedges, and Duncan (1991). In this model, a double encoding of spatial location (fine-grained and categorical) is assumed. The **fine-grained encoding** is conceptualized in terms of coordinates, the **categorical encoding** in terms of a region with a prototypical value and boundary values. While memory is supposed to be inexact, recall of a spatial location is the consequence of a weighting of the two memory traces. Category information is used to estimate the position of the Figure. The weighting of the two encodings occurs on the basis of the (inexact) coordinate values and the prototype of the remembered category, which leads to a bias of the reproduction reports towards the **prototype**. A second source of bias in the Huttenlocher *et al.* model is **bias away from boundaries**. This bias results because fine-grained encoding close to category boundaries will be reproduced in such a way that the (inexact) fine-grained encoding complies with the remembered category encoding. Interpreting the results with this model lead Crawford *et al.* to conclude that prototypes of spatial categories exist in between the axes. The boundaries on the spatial level (i.e. the axes) are reflected in spatial language as prototypes,

---

[1] Linguistic as well as non-linguistic data are thus obtained in a single experiment. The possibility that the rating influences the subsequent reproduction cannot be excluded. Because the two kinds of data seem to oppose one another, Crawford *et al.* conclude that there is no such influence. The results of the rating task show highest applicability of *above* on the vertical axis extending upwards, with applicability diminishing with distance from the axis. This result is similar to a rating experiment described in Hayward and Tarr (1995).

hence an inverse relation between spatial language and spatial representation is suggested. According to Crawford *et al.* a single structure may underlie categorization in non-linguistic and linguistic spatial tasks; the structure plays a different role in the two tasks, however.

Crawford *et al.* do not consider the two kinds of bias separately. Bias away from category boundaries and bias resulting from the weighting of the two memory traces point in the same direction and are considered as a whole. Whether the observed bias is a bias toward prototypes, a bias away from boundaries, or both, remains unclear. Implicitly four mutually exclusive regions are assumed. But, although, for example, *above* is opposed to *below* and mutually exclusive regions may be necessary at the non-linguistic level to account for this, *above* is not necessarily opposed to *to the left of*. Thus, the regions that *above* and *to the left of* refer to may be partially overlapping. *Left* vs. *right* and *above* vs. *below* may refer to two pairs of mutually exclusive **half-planes** oriented in an orthogonal manner. The bias away from axes that Crawford *et al.* found may have been caused by boundaries and not by prototypes.

Huttenlocher *et al.*'s assumption of two kinds of memory traces in spatial relation reproduction tasks is intuitively appealing. When a participant is asked to reproduce the position of a Figure as exactly as possible, it does not suffice for the participant to remember that the Figure was *in front of* the Referent (spatially and/or linguistically), but also where *in front of* the Referent the Figure is located. In the experiments of Crawford *et al.* the reproduction of the Figure in positions on the horizontal and vertical axes is more accurate than in other locations. Contrary to Hayward and Tarr (1995) enhanced accuracy is associated with category boundaries and not with prototypes. The approach explained below will also hold that axes are represented as boundaries. However, the axes bound half-planes and not quadrants.

1.3. Axes as Boundaries, Half-Axes as Best Examples

As shown above different views exist regarding the correspondence between the representation of directional spatial relations and the linguistic reference to those relations. Here, an alternative conception for intrinsic object-centered directional relations is proposed. This conception may also resolve some of the apparent conflicts in the interpretations of results from experimental situations with overlapping frames of reference. The present view holds that—in the 2-D object-centered situation—intrinsic axes are boundaries between pairs of regions and **half-axes** serve as the best examples of regions. This view is illustrated in more detail by considering how an intrinsic object-centered frame of reference might be imposed on a Referent. It is assumed that establishing an intrinsic frame of reference is an active process requiring **attention** (Logan, 1995) and that the frame has to be extracted from the Referent's structure. The idea of extraction is important in relation to the experiments described in section 2 below. The experimental task involves reproduction of an intrinsic spatial relation while the Referent changes its orientation between encoding and reproduction of the spatial relation. It could be argued that such a task is only possible if a frame of reference is extracted during the encoding of the spatial relation (in order to remember the Figure's location relative to the Referent's intrinsic structure) as well as during the reproduction of the spatial relation (in order to apply the remembered information to the Referent's changed orientation).

Figure 12.2 A1 shows a possible Referent (the oval shape) with a Figure (the small circle).[2] On the basis of the Referent's features the Figure's location has to be assigned to regional categories and therefore the space surrounding the Referent has to be divided up. The only features of this Referent are shape features. An interpretation of the shape structure has to be extended outwards into regions around the Referent. Two axes can be defined on the Referent, one being an axis of symmetry (axis a) and the other oriented orthogonally to this axis (axis b). See Fig. 12.2 A2. It is assumed here that (reflection) symmetry is a strong perceptual grouping principle (see e.g. Wagemans, 1995) and that axis a is represented on the perceptual object as a result of perceptual processing. The status of axis b is less clear. Here, axis b is positioned in such a way that it intersects the contour at its greatest width. Other positionings are possible: Axis b might cross axis a at the Referent's center of gravity or at axis a's midpoint. Axis b may be represented less exactly than suggested in Fig. 12.2. Boundaries and prototypes in the model of Huttenlocher, Hedges, and Duncan (1991) have an inexactness value. Axis b might have one as well. Finally, axis b may not be represented at all yet, and may instead be constructed as a result of distinguishing (separating) object parts as described below.

No directional information is available at this point. The presence of an **axis of symmetry** only signals that the Referent is reflectionally invariant about axis a. It is important to acknowledge the difference between axes (dimensions) and the sign attached to half-axes (directions; see Landau (Ch. 2) for some interesting observations regarding the independence of the representation of axes and the representation of directionality assigned to axes). Using axis b as a boundary, the two object parts separated by this axis can be differ-

---

[2] Harris and Strommen (1974) used a similar object and asked participants to draw a line in the object separating the front from the back. The typical result (over 80 per cent of the responses) is that the line separates the 'tapered' end from the 'broader' end (as axis b in Fig. 12.2 does).

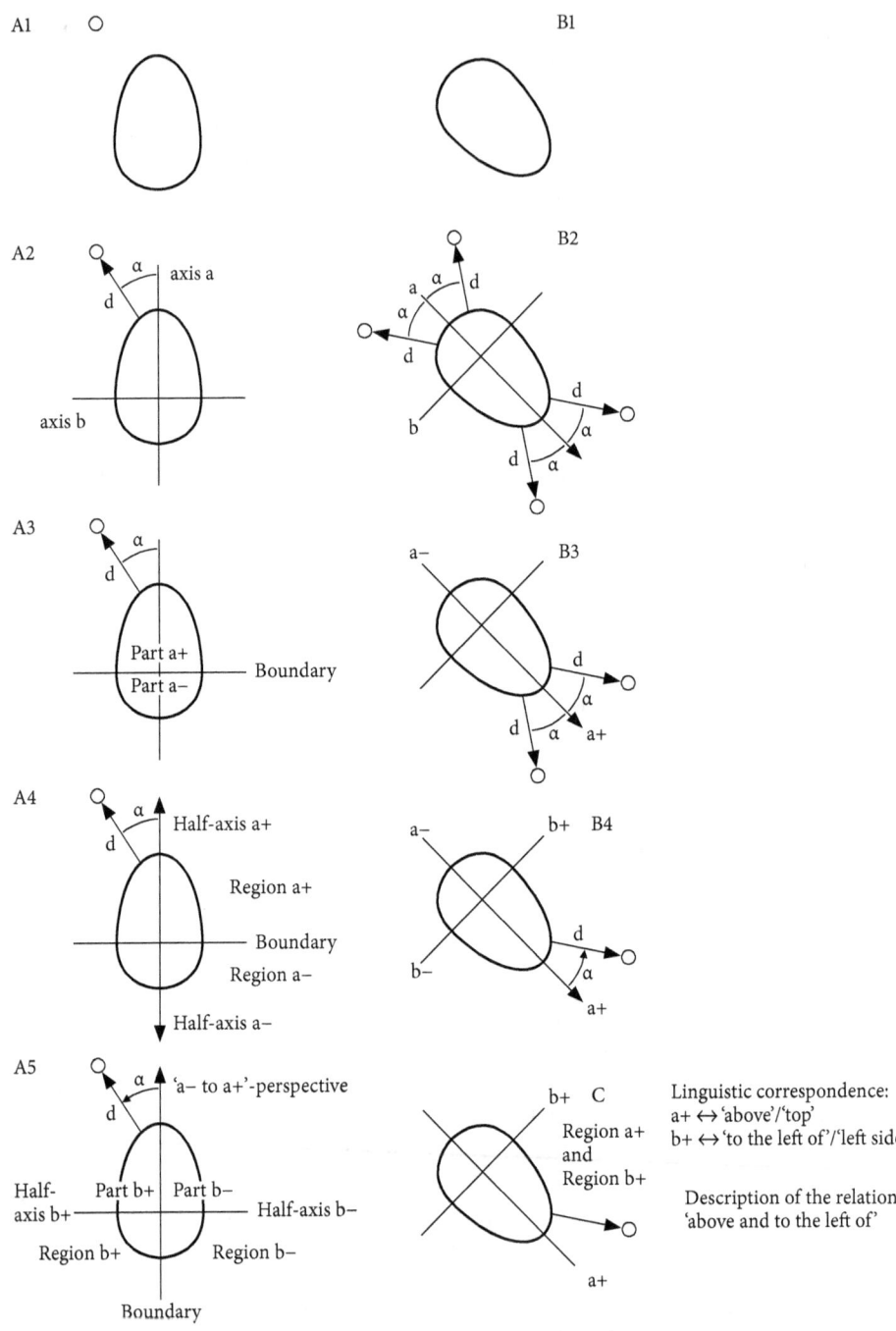

FIG. 12.2 Encoding and reproduction of an intrinsic spatial relation.

entiated by their shape (Part a+ vs. Part a−). See Fig. 12.2 A3. Two half-planes as well as two half-axes constituting axis a are also separated by axis b and can be associated with the differently shaped object parts. When the part difference is extended to the half-planes, one pair of regions can be distinguished (Region a+ vs. Region a−). When the part difference is extended to the half-axes constituting axis a, these half-axes can be distinguished (Half-axis a+ vs. Half-axis a−), making axis a directed. See Fig. 12.2 A4. Importantly, it has to be established first that object parts are different before the difference can be extended to regions outside the object and to half-axes, because intrinsic reference has to be based on the Referent's structure. Intrinsic reference is only possible when distinctions within the Referent's structure can be made. In this example the distinctions are within the shape structure. Making spatial distinctions, of course, precedes categorization and linguistic labeling.

Representing axis a as a boundary in the same way as was done with axis b above does not suffice to distinguish another pair of regions and half-axes. The **mirror symmetry** of the two object parts separated by axis a allows no direct discrimination between them (i.e. they have the same shape). However, the acquired directionality of axis a makes it possible to distinguish the object parts, regions, and half-axes on either side of axis a. The explicit interpretation of the symmetry axis as, e.g., the 'a− to a+'-perspective, allows for the indexing of one part separated by axis a as the b+ part and the opposite part as the b− part and a similar indexing of the half-planes and half-axes as well. In this way axis b becomes directed. See Fig. 12.2 A5. Note that in this case the object parts do not necessarily have to be differentiated before the regions and half-axes can be distinguished (all distinctions can be based on the 'a− to a+'-perspective).

In Fig. 12.2 the fine-grained encoding of the relation between Figure and Referent is defined with a **vector**. This vector is based on the closest distance between the objects. This definition of fine-grained vector information suffices for the present example. However, vectors may be defined in other ways (see Carlson, Regier, and Covey, Ch. 6). Within the present context, the vector has to be linked to the intrinsic reference frame, so that the location of the Figure can be interpreted relative to the Referent's structure. This may be done by decomposing the vector into an **orientation** (α) and a **distance** (**d**). The orientation is defined relative to one of the Referent's (half-) axes while the (closest) distance between Figure and Referent has to be defined relative to some aspect of the Referent's extendedness.

In the process of reference frame extraction the orientation of the vector may be linked consecutively to, respectively, axis a, half-axis a+ and finally to half-axis b+ and/or region b+ as well. A spatial representation may then result which links the specific location of the Figure to the reference frame of the Referent and which may be described with, for example, the Figure is at distance d, oriented α away from half-axis a+ and toward half-axis b+ (or in region b+). Progressively linking the vector during reference frame extraction may also occur during the recall of an intrinsic position. The Referent is shown in a different orientation without the Figure in Fig. 12.2 B1. If axis a and axis b are represented (alternatively, if only axis a is defined) four possible vectors with α and d can be specified. See Fig. 12.2 B2. When the a+ and a− parts and—thereafter—the a+ and a− regions and/or half-axes are differentiated, two possible vectors can still be defined (Fig. 12.2 B3). If, finally, the b+ and b− half-axes and/or regions are distinguished, the vector and therefore the former position of the Figure can be determined unambiguously (Fig. 12.2 B4).

Assuming that the above-mentioned spatial distinctions constitute the information that is represented for intrinsic object-centered frames of reference, the remaining question is how the spatial representation corresponds to directional prepositions. The correspondence can be relatively simple: a+ vs. a− might correspond to, e.g., the *above* vs. *below* dimension, and b+ vs. b− to the *left* vs. *right* dimension. The **regions** and **parts** can be referred to with the corresponding terms (e.g. *above* for a region and *the top* for a part). The Figure (and the vector) in Fig. 12.2 C is in region a+ and in region b+. Therefore, the relation can be described with two prepositions such as *above* and *to the left of*. If the Figure's location is on an axis (the vector coincides with the axis), say on half-axis a+, then only *above* will be used because axis a is neutral with respect to the other directional opposition (axis a is neither in region b+ nor in region b−). This has as a consequence that a spatial relation in which a Figure's location can be characterized with a vector that coincides with a half-axis is the best example of a directional preposition. In spatial relation description experiments the question is not necessarily where, for example, *above* ends and *to the left of* begins. Instead, the question might be where a single directional preposition or where a combination of directional prepositions is used (for a similar way of analyzing data see Zimmer, Speiser, and Baus, 2001).

The present view has some commonalties and some differences with the views of Hayward and Tarr (1995), and Crawford, Regier, and Huttenlocher (2000). Axes do function as boundaries (as in Crawford *et al.*), but as boundaries between half-planes, and not between quadrants. Half-axes do have a function as well in the correspondence between the representation of directional spatial relations and linguistic reference to such relations. Positions on half-axes (vectors coinciding with half-axes) are the best examples of directional prepositions. These positions are categorized only in terms of one of two spatial dichotomies.

Using axes as boundaries to establish spatial dichotomies is an example of a feature model. Being located on one or the other side of an axis is necessary and sufficient for categorization. At first glance this seems in sharp contrast to Hayward and Tarr who propose a purely prototypical model in which goodness-of-fit measures determine categorization (see Talmy (1983) and Hayward and Tarr (1995) for a discussion of prototype and feature models in the present domain). However, the featural aspect of the present model applies to categorization within a dimension (categorizing a relation as one direction or the opposing direction). Therefore the present approach does not exclude that goodness-of-fit measures are used to determine if a specific relation matches more with one half-axis or with a half-axis of another dimension. Prototype-like effects in linguistic behavior may thus occur when the orientation of a vector is compared with two half-axes.

1.4. Axes, Object Features, and Reference Frame Extraction

The extraction of a frame of reference as described above has parallels with the **intrinsic computation analysis** of Bryant and Tversky (1999). In their experiments participants learn the locations of six objects positioned in relation to axial directions of a model of a person. After a configuration has been learned the participant is told that the person in the configuration changes his position and faces another object. The participant is then probed with an axial direction term (*head, feet, front, back, left,* or *right*) and has to decide as quickly as possible which of the objects is in that direction with respect to the person in the scene. If the participant imagines to be looking at the remembered configuration the intrinsic object-centered frame of reference has to be extracted from the imagined person (alternatively, a participant may imagine taking the perspective of the person, in which case an intrinsic ego-centered frame of reference would be used). Responses are found to be fastest on the *above–below* dimension, followed by the *front–back* dimension, and, finally, the *left–right* dimension, independent of the orientation of the imagined person. See Corballis and Cullen (1986) and Logan (1995) for similar results obtained with perceptually available stimuli.

This pattern of reaction times is explained with the intrinsic computation analysis. According to this analysis of spatial relations in the object-centered frame of reference, the *above–below* axis is defined first in order to identify the orientation of the object, then the *front–back* axis, and finally the *left–right* axis, which cannot be defined unless the other two axes are already defined. Differences in the accessibility of reference axes are thought to be caused primarily by the axes' relative salience that depends upon their respective symmetries (Tversky, 1996). Tversky refers to the front–back and the top–down axes of a body as **asymmetric axes**, and the left–right axis as a **symmetric axis**. Similarly, Landau and Jackendoff (1993) refer to directed and symmetric axes respectively. In Landau and Jackendoff's terminology, axis a in Fig. 12.2 is a directed (asymmetric) axis because there are 'inherent regularities that distinguish one end from another', and axis b is a symmetric axis because there are 'equivalent elaborations at both ends of the axis'. Geometrically, axis a is an axis of symmetry. In Tversky's and Landau and Jackendoff's use of the terms, axis b is a symmetric axis by virtue of the whole object being bilaterally symmetric over axis a. According to Tversky (1996), asymmetric axes are relatively easily accessible.

One parallel between the intrinsic computation analysis and the present view is that the spatial distinction b+ vs. b− cannot be made unless the distinction a+ vs. a− is already made. Although the emphasis is on regions here, and not on axes, the principle is similar. Another parallel between both approaches is that there may be differences in the accessibility of regions due to the relative (a)symmetry of the object parts. If the present view is extrapolated to the 3-D case, then an axial plane may separate different parts (say the front and the back). The axis associated with and intersecting these distinguishable parts is categorized as the front–back axis (i.e. its half-axes are defined and the axis becomes directed). At the same time, the regions *in front of* vs. *behind* are easily accessible compared to the regions *to the left of* and *to the right of* whose accessibility requires that the directionality of the other axes is already established. In the 2-D situation, the accessibility of regions depends on the relative salience of the difference (asymmetry) between the parts on either side of the axial boundaries.

Similar principles of relative (a)symmetry are derived by van der Zee and Eshuis (Ch. 11) in order to explain the categorization of axes in novel objects. In one of the experiments described there, participants indicated the intrinsic regions *in front of* and *behind* or *to the left of* and *to the right of* around abstract objects. Van der Zee and Eshuis find that axes along which **contour expansion** occurs and axes oriented orthogonally to a **curved axial plane** are most **directionally marked**. According to their **Spatial Feature Categorization model** an axis that is most directionally marked is categorized as the above–below axis, the intermediate directionally marked axis is the front–back axis, and the least directionally marked axis is the left–right axis. The axes that are most directionally marked as evidenced by their categorization must obtain this marking from spatial features. Interestingly, the axes mentioned have something in common: they intersect two different (and thus distinguishable) sides. An axis along which contour expansion occurs intersects a relatively small and a relatively big side. An axis oriented orthogonally to a curved axial plane intersects a **convex** and a **concave** side.

Extrapolating from the findings of Bryant and Tversky and van der Zee and Eshuis it can be expected that the extraction of an intrinsic frame of reference takes place according to the order of 'directional marking', or, to put it differently, the relative distinctiveness of the object parts an axis intersects (or, to describe it in line with the last section, the relative distinctiveness of the object parts the orthogonal axis separates). In what follows, the extraction of regions will simply be assumed to follow the *top/bottom, front/back, left/right* order based on linguistic data that have already been obtained with the objects that are used in the spatial memory task that is described below. This hypothesis will be referred to as the **regional interpretation of the intrinsic computation analysis.**

2. Experiments

In the experiments participants have to reproduce spatial relations by putting a Figure (a dot) back on the location it was before, relative to a Referent. The use of information in the scene other than information based upon intrinsic characteristics of the Referent is made ineffective by changing the orientation of the Referent between the encoding of the spatial relation and the subsequent reproduction of this relation. In this way the intrinsic object-centered frame of reference can be investigated regarding its effects on reproduction performance. Participants are explicitly instructed to remember and reproduce the location of the dot relative to the Referent's perspective. For a more extensive description of the experiments, see Eshuis (forthcoming).

2.1. Experiment 1: On-axis versus off-axis Locations

In the first experiment two Referents are used. One of these is very similar to one of the Referents used by Hayward and Tarr (1995) and the other is very similar to a cross-section of one of the Referents used in the experiments described in van der Zee and Eshuis (Ch. 11). These objects are used in the experiment because linguistic data regarding their frames of reference has already been collected. The frames of reference observed by van der Zee and Eshuis and by Hayward and Tarr are used to define the off-axis and on-axis locations in the present experiment. The Referents with their intrinsic directions are illustrated in their canonical orientations in Fig. 12.3. Three remarks have to be made at this point. First, there is an important difference in the way the frames of reference of the objects were determined by Hayward and Tarr and by van der Zee and Eshuis. In the Hayward and Tarr experiments participants had to produce a description of a given spatial relation, while in the van der Zee and Eshuis experiments participants had to produce spatial relations on the basis of given descriptions. It is assumed here that the intrinsic frames of reference found with these different tasks can be treated alike. Second, there is no independent linguistic data regarding the intrinsic frame of the Referent in the Hayward and Tarr experiments, because several overlapping frames of reference were active. It is assumed here that the data found with overlapping frames of reference can be extrapolated to the intrinsic reference situation. Finally, in experiments 1 and 2, the participants are native speakers of German, while the participants in the experiments of van der Zee and Eshuis were native speakers of Dutch, and the participants in the experiments of Hayward and Tarr were native speakers of English. It is assumed here that there is no important difference between (native speakers of) these languages regarding the representation of intrinsic directional relations.

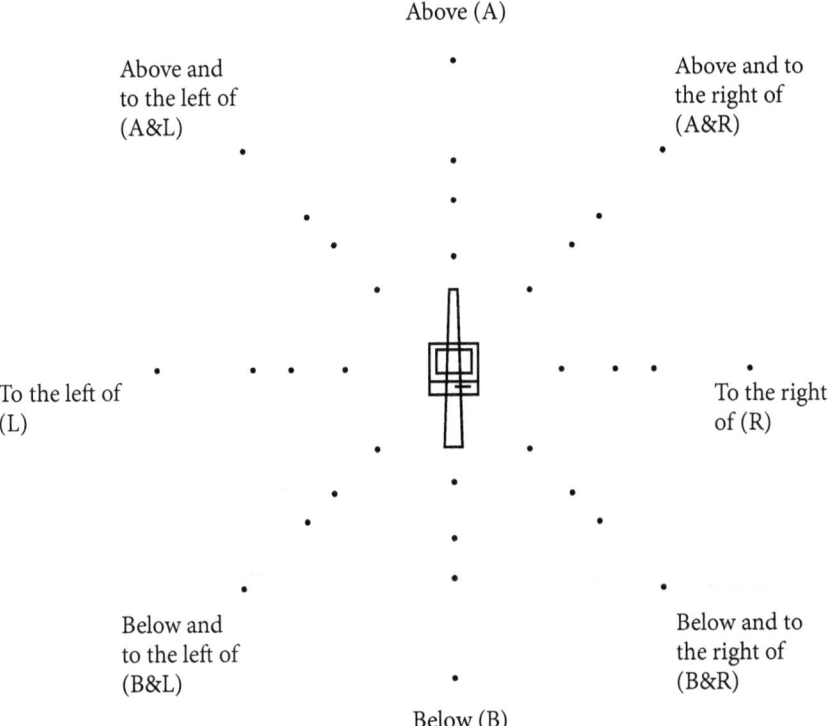

FIG. 12.3 The two Referents used in experiment 1. Black dots represent the test locations. The labels illustrate the linguistic reference to the directions for both Referents as reported by Hayward and Tarr (1995) and van der Zee and Eshuis (Ch. 11).

In the experiments reported here participants reproduce spatial relations by putting a dot back to its former location with respect to a Referent. Between the encoding of the relation and its reproduction the Referent's orientation changed 0, 45, 90, or 135 degrees. Eight directions are defined around the objects (4 on-axis and 4 off-axis), and four distances in each direction (32 dot locations per object, see Fig. 12.3). Participants have to perform a double task in the reproduction phase: first, they determine visually the previous position of the dot relative to the Referent as quickly as possible and press a key. Following that, they indicate this position as accurately as possible with the mouse cursor. Respectively, reaction times as well as reproduction coordinates are measured.

Reproduction on-axis is expected to be more accurate than reproduction off-axis. The reaction times are expected to correspond to the predictions made by the regional interpretation of the intrinsic computation analysis. Reaction times are thus expected to be fastest on the *above* and *below* half-axes, followed by the *left–right* regions (locations in the overlapping parts of, e.g., *to the left of* and *in front of* can be determined unambiguously only when the *left–right* dimension is extracted, therefore only locations on the *above* and *below* half-axes should be faster).

Figure 12.4 illustrates how estimates of dot location are treated in the construction of dependent variables. Reaction times co-occurring with reproduction estimates that are not in the 'Correct' area as defined in Fig. 12.4 were excluded from the analysis as well as extreme reaction times. Figure 12.5 shows the mean reaction times for each object and intrinsic direction. Figure 12.6 shows the standard ellipses of the estimated Figure locations for each true location relative to the objects. Regional error as defined in Fig. 12.4 was not observed in the spatial relation reproduction experiments mentioned in section 2. The errors found here can be considered categorical: Instead of, for example, region b+ (*to the left of*), region b– (*to the right of*) has been chosen during reproduction, while fine-grained information does not seem to be impaired more for dots reproduced in an 'Incorrect' region compared to dots reproduced in the 'Correct' region (as can be seen in Fig. 12.4, the same local variability is allowed for estimates in the 'Correct' region and in the 'Incorrect' regions). Table 12.1 shows percentage regional error for each object together with the amount of error that was not captured by any of the regional error categories.[3]

---

[3] The procedure assures that the a priori chances that an estimate will be considered 'Correct' or to be any of the regional errors are equal. It also assures that similar percentages of estimates within the 90° area around the true location are considered 'Correct.' However, the a priori chances that an estimate will be considered 'Correct' or a regional error as opposed to 'Other' are different for each intrinsic location.

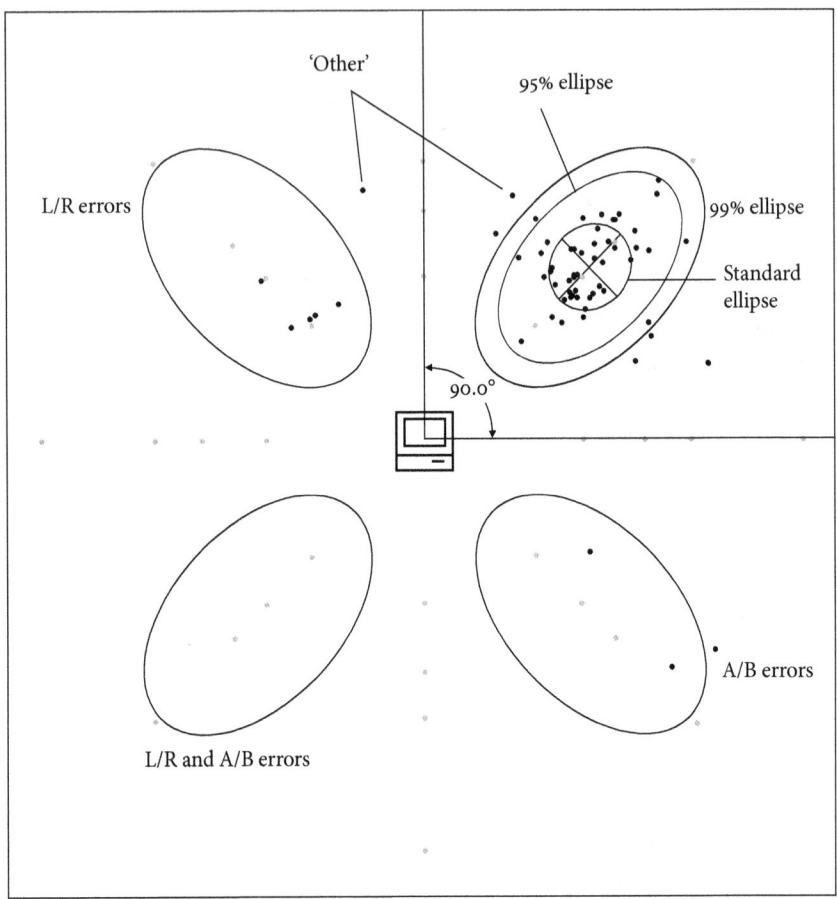

FIG. 12.4 Illustration of the treatment of Figure location estimates. The estimates within the correct 90° area around the true location are used to compute an ellipse that fits 99% of the estimates (assuming a Cartesian bivariate normal distribution). The scores within this ellipse are considered to be 'Correct.' Mirror images of the ellipse are made to define areas for regional 'Incorrect' estimates. Estimates within these ellipses are 'Left–Right' (L/R), 'Above–Below' (A/B), or 'Left–Right and Above–Below' (L/R and A/B) errors. (For on-axis true locations only one mirror image is made.) The 95% ellipse in the 90° area functions as a filter to remove extreme scores. Only estimates within this ellipse are used for the analysis of reproduction accuracy. The standard ellipse based on these estimates is drawn. Standard ellipses are a descriptive tool (a kind of two-dimensional standard deviation) and contain about 40% of the estimates with which they were defined.

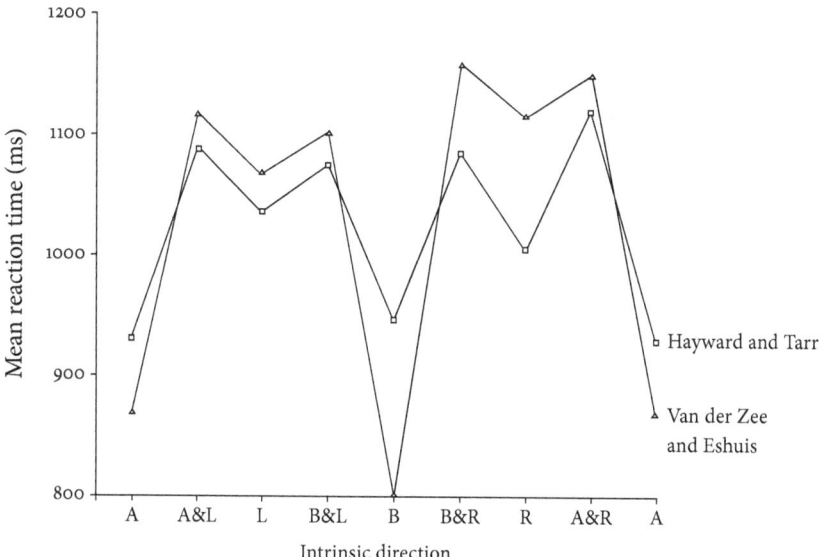

FIG. 12.5 Reaction times for both objects and each direction. Abbreviations as defined in Fig. 12.3. The above direction is presented twice.

As can be seen from the figures some differences exist between the two Referents, although both follow more or less the same pattern. For both objects the above–below direction is faster than the left–right direction, which is faster than the off-axis directions. This finding is in line with the intrinsic computation analysis for the on-axis directions. The slower response on the off-axis directions is in line with the regional interpretation in that the left–right regions show slower reaction times than the on-axis above–below directions. That responses on the off-axis locations are slower than responses on the on-axis left–right locations was not expected but is not necessarily contradicting a regional pattern. The off-axis locations are in two regions, and this may lead to a somewhat higher memory load if memory is based on regions: off-axis locations require that values on two directional oppositions are remembered (say, a+ and b+ corresponding to *above* and *to the left of*) while on-axis locations only require one directional opposition (say, b+ corresponding to *to the left of*). Alternatively, the reproduction of off-axis locations may demand an additional orienting of a vector relative to the axes.[4]

---

[4] These findings may not be solely attributable to reference frame extraction during the reproduction phase of a trial. A frame of reference has to be extracted during the encoding phase as

TABLE 12.1 *Error types for each Referent*

Referent	Direction	Cases	Other error (%)	Correct (%)	L/R error (%)	A/B error (%)	A/B and L/R-error (%)
Hayward and Tarr	Above/below	All	4.5	93.7	n/a	1.7	n/a
		Ellipses		98.2		1.8	
	Off-axis	All	7.9	85.1	4.7	1.3	1.0
		Ellipses		92.5	5.1	1.4	1.0
	Left/right	All	6.5	89.5	4.0	n/a	n/a
		Ellipses		95.7	4.3		
van der Zee and Eshuis	Above/below	All	2.1	97.9	n/a	0.0	n/a
		Ellipses		100.0		0.0	
	Off-axis	All	8.9	82.6	7.0	1.1	0.4
		Ellipses		90.7	7.7	1.2	0.5
	Left/right	All	6.4	85.6	8.0	n/a	n/a
		Ellipses		91.4	8.6		

Note: Abbreviations are as defined in Fig. 12.4. The first line for each Referent's direction grouping shows all estimates; the second line shows only the estimates in the (equal-sized) ellipsoid areas.

Regarding the percentage regional errors the pattern follows (more or less) the reaction time pattern. Directions where relatively more regional errors are made are directions that show relatively longer reaction times. No hypotheses have been defined for regional error. One might assume that the explanations for reaction time differences are identical to those for regional errors.[5] However, this is not necessarily so. The majority of regional errors are left–right errors. This questions whether reference frame extraction effects alone can be used to explain left–right uncertainty.

well. Insufficient extraction of the reference frame during encoding may result in an incomplete memory trace for the Figure's location. This may then lead to uncertainty during the reproduction phase that may produce longer reaction times. However, reference frame extraction is the same for both phases. Because of the restricted encoding time, relative occurrence of insufficient encoding for a particular dimension should follow the order in which the dimensions are extracted. The findings may thus show an exaggerated extraction pattern.

[5] See n. 4 and substitute 'regional error' for 'longer reaction times.'

Accuracy of reproduction is better on-axis than off-axis (whether one considers vertical and horizontal error separately or considers only angular error), and for the abstract object there exists a difference between the above–below and the left–right axis. The experiment reproduces the reproduction accuracy results of Hayward and Tarr (1995) in the intrinsic object-centered frame of reference regarding the Hayward and Tarr Referent. The difference in reproduction accuracy between the above–below and the left–right on-axis locations of the abstract Referent seems to question the generality of the finding. However, a variety of reasons can be thought of to account for this difference: for example, the assumed position of the left–right axis may have been wrong, or the above–below axis may be represented more exactly than the left–right axis.

For present purposes, the regional error is most interesting. This kind of error may also reveal something about how spatial information characterizing directional relations is represented. There are now two measures with which the correspondence between spatial representation and spatial language can be investigated. On the one hand, directional prepositions may correspond to axial prototypes that have an influence on (local) accuracy of reproduction. If this is the case the above–below axis produces similar results as the left–right axis. This correspondence is the one suggested by Hayward and Tarr: axial prototypes exert an influence in linguistic as well as non-linguistic spatial tasks. On the other hand, the correspondence between spatial representations and language can also be investigated with the amount of regional error. In that case there is a difference between the above–below and the left–right dimension. Errors between the left and the right regions occur relatively frequently.

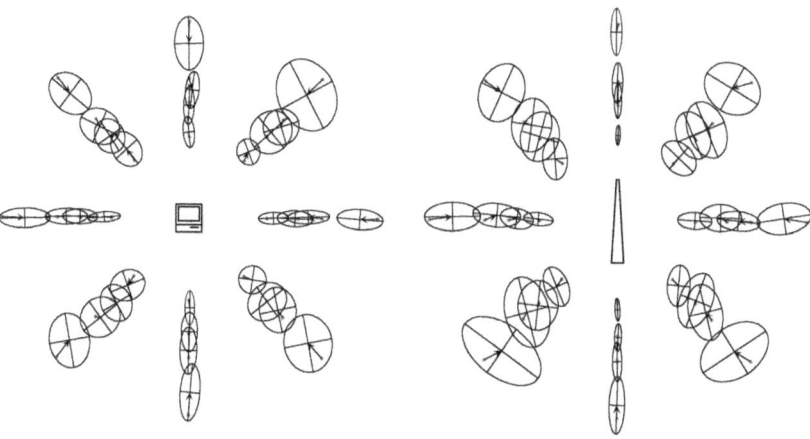

FIG. 12.6 Standard ellipses for each location relative to the Referents. Arrows denote the bias from the true location to the mean of reports.

Errors between the above and the below regions happen less often. Errors between above and below regions and left and right regions at the same time are very rare. Regional errors are thus typically made from a region at one side of an axial boundary to the region at the opposite side of the boundary. The two pairs of half-planar regions are susceptible to regional error to a different degree. This is in line with the 'axis as boundary' point of view as well as an ordered extraction of regions. The regional interpretation of the reaction time data seems to support this option.

As mentioned before, the majority of the regional errors are left–right errors. On the one hand, this may indicate that participants just had difficulties with symmetry. On the other hand, the difference in frequency of the kinds of regional error may point to the relative independence of the respective dimensions. Moreover, symmetrical properties of a Referent may determine how the Referent is spatially structured. Extrapolating from that structure or from axes oriented in line with the structure, such properties may determine how the surrounding space is carved up. The next experiment explores local reproduction accuracy and regional performance with an asymmetric Referent in order to see if regional error is merely caused by the symmetry of the Referent.

2.2. Experiment 2: Exploring Local and Regional Error

In the second experiment, memory for many intrinsic locations around a Referent is investigated. The Referent is curved, has an expanding contour, and is very similar to a cross-section of a Referent used by van der Zee and Eshuis (Ch. 11). Around the Referent 72 intrinsic directions and three distances in each direction are defined. There are various reasons for using this Referent and defining many more test locations around it. On the one hand, it is not clear whether the intrinsic axes of the Referent are used or if the axes of the bounding box are used. Which of these two axial systems reported in van der Zee and Eshuis is used in the intrinsic relation reproduction task might become clear through the use of a substantial number of test locations. Many test locations make it possible to consider the pattern of relative accuracy of reproduction more closely, whether the pattern shows an axial effect, and, if so, to which axial system it can be ascribed. Figure 12.7 shows the Referent, the two possible axial systems with the preferred linguistic labeling according to van der Zee and Eshuis, and the test locations.

Another reason to use this Referent is that it is a non-symmetric object. If the regional errors observed in experiment 1 merely reflect difficulties of participants with the Referent's symmetry, the occurrence of regional error should be much smaller here. However, if with this Referent regional errors are still found, one may consider whether there is a correspondence between the preferred categorization of the axes as reported in van der Zee and Eshuis and the amount of regional error made over each of the axes. It has been hypothesized that the preferred categorization is caused by the relative distinctiveness of object parts. This relative salience of differences between object parts should have an effect on the order of region extraction, which may influence the amount of regional errors that are made. Although the preferred categorization of the axes of this object is similar to the categorization of the axes of the abstract Referent in the first experiment, the intersubject-agreement on axis categorization with this object is lower. This is in line with the conclusion in van der Zee and Eshuis that an axis orthogonal to a curved axial plane is relatively more directionally marked. The difference in directional marking of

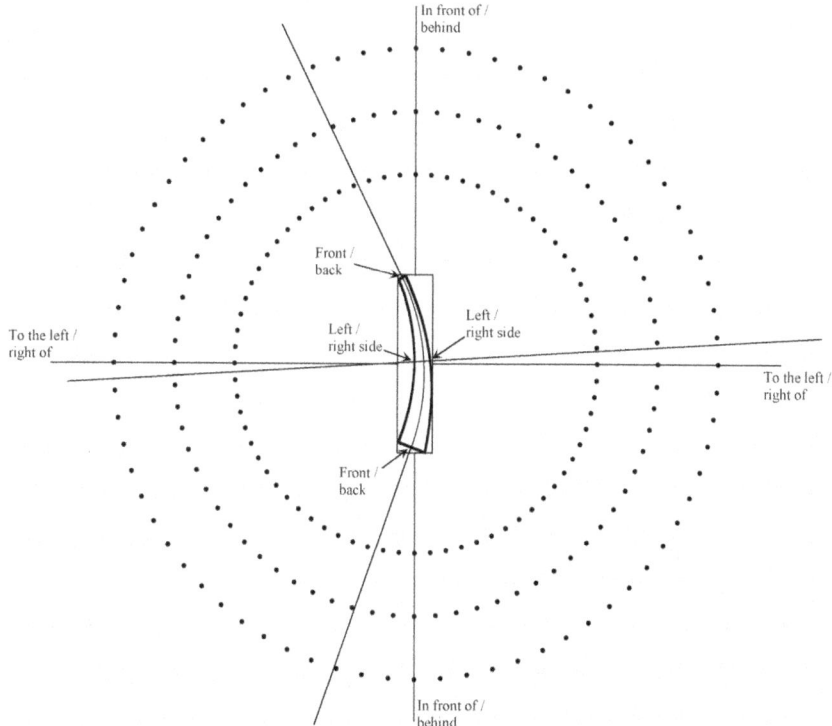

FIG. 12.7 The Referent used in experiment 2. The black dots denote the test locations. Two possible axial systems are drawn: one represents the intrinsic axial system of the Referent, the other represents the axial system of the bounding box. Categorization of the axes as reported by van der Zee and Eshuis (Ch. 11).

the axes of the present Referent (whichever axial system is being used) should therefore be smaller than the difference in directional marking of the axes of the abstract object in experiment 1. It is thus expected that the difference in the amount of regional error for each of the dimensions will be smaller as well.

Participants reproduced dot locations relative to the Referent. The Referent's orientation changed 30, 60, 90, 120, 150, or 180 degrees between the encoding of the relation and its reproduction. In this experiment only reproduction coordinates were measured. After performing the experiment participants were asked how they had remembered where the location of the Figure was.[6] Interestingly, the answers to the question, 'In what way did you remember the position of the dot?' seemed to separate into 'categorical' and 'metrical' strategies. A categorical strategy is defined here by participants mentioning that they remembered a part of the Referent in order to perform the task. A metrical strategy is defined by participants mentioning that they imagined using lines and/or angles. Most participants mentioned both kinds more than once. The categorical answers were always one of two paired concepts, although they were described differently by the participants. One pair was mentioned by 74 per cent of the participants. This opposition was usually referred to with *innen–außen* ('inside-outside'), but also with *unten–oben* ('under–above') and *hohl–konvex* ('hollow(concave)–convex'). With these terms the participants referred to the curved sides of the Referent and/or regions adjacent to them. This opposition will be referred to as concave–convex below. The other categorical opposition was mentioned by 70 per cent of the participants and was referred to as *dünn–dick* ('thin–thick'), *klein–groß* ('small–big'), *eng–weit* ('narrow–wide'), *oben–unten* ('above–below'), and *vorne–hinten* ('front–behind (back)'). With these terms the participants referred to the two straight sides and/or regions adjacent to them. This opposition will be referred to as thin–thick below. The metrical strategies mentioned most frequently by the participants were that they used—or imagined—extensions of the curved sides (86%), extensions of the straight sides (65 per cent), and triangles between two corners of the Referent and the location of the Figure (52%) to help them in the reproduction of the dot location.

To illustrate which axial system the participants were using standard ellipses are defined in the same way as described in Fig. 12.4. This procedure is quite neutral with respect to the possible axial systems. The results of this calcula-

---

[6] Spatial (non-linguistic) data and linguistic data are thus collected in a single experiment. For the spatial data, this should have no effect because participants were not told in advance that they would be asked for their strategies after the experiment. The labeling of object parts after the experiment does not necessarily mean that the labels were used by the participants during the experiment (although this is possible).

Memory for Locations Relative to Objects 249

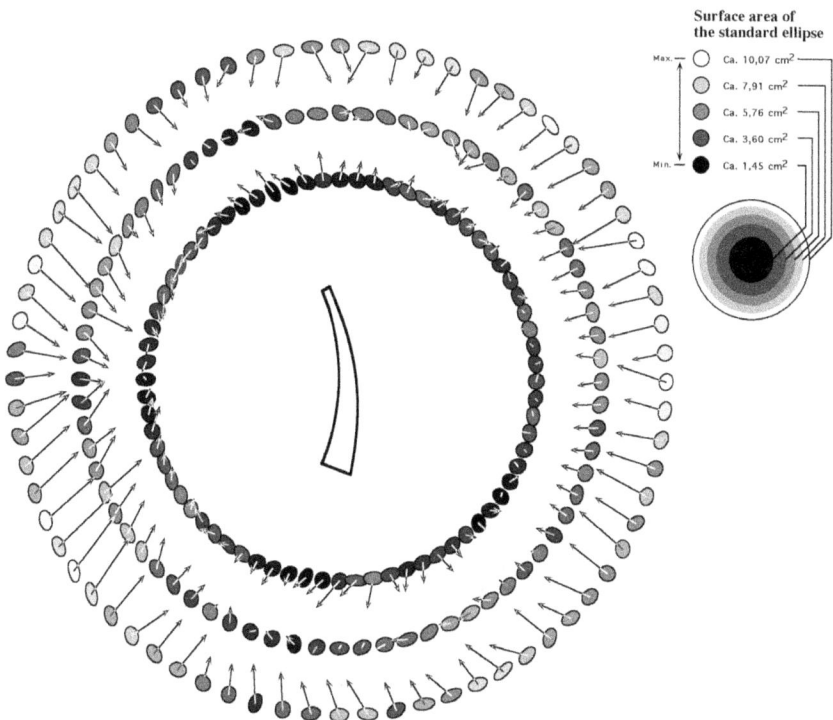

FIG. 12.8 Standard ellipses for each location around the Referent. All ellipses are presented with a standard surface area and centered at the true locations for illustrative purposes. The aspect ratio and the orientation of the ellipses have been preserved. The actual situation involves standard ellipses with variable surface areas (differentiated by the shades of gray) while they are centered at the tips of the arrows (which denote bias).

tion are shown in Fig. 12.8. As can be seen from the figure, if an axial system is revealed by the accuracy of reports, it is the intrinsic axial system of the object (and not the axial system of a bounding box). A few important qualifications have to be made. There is better reproduction performance around the area of the curved axis extending outward at both its ends. The same can be seen for the orthogonal axis when its end intersecting the concave side is extended outward. However, extension of this axis from its end intersecting the convex side does not show an enhancement in reproduction performance.[7] Note

---

[7] Not finding higher reproduction accuracy around the convex axial direction of the Referent could be explained in different ways: There is no axis, an axis does not necessarily lead to better reproduction performance, there is only an inexact axis, and/or there is no intersubject agreement on the exact location of the axis.

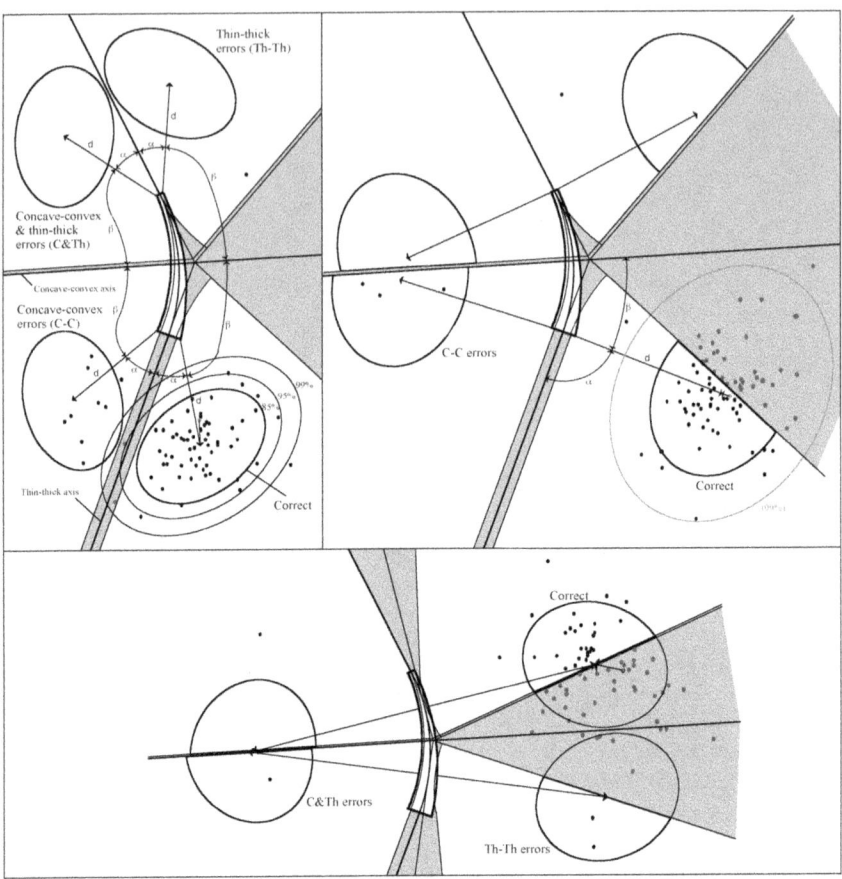

FIG. 12.9 Construction of the regional error measure in experiment 2. Top left: Location estimates that are in the same area between two half-axes as the true location are used to fit ellipsoid areas. Often, however, ellipses fitting 99% of the estimates would extend over a half-axis. Ellipses fitting 85% of the estimates are used here to allow a reasonable amount of test locations to be considered while at the same time true locations close to axes are excluded (the latter locations do not seem to allow a clear separation between 'Correct' and regional 'Incorrect' responses). Corresponding regional error ellipses are constructed whereby proximal distance to the contour (d) as well as orientation to the half-axes α and β are preserved (note that orientation is measured along the Referent's contour). Top right: Regional error ellipses constructed like this may also extend over half-axes. Therefore only those parts of the ellipses for which a definition in all four areas is possible are used to estimate regional error. This excludes estimates in gray areas. Bottom: Gray areas may differ depending on the mean of the estimates that are used to calculate orientation and proximal distance.

that the accuracy of reproduction may be in line with the metrical reports of the participants. The often-mentioned metrical strategy of extension of the straight sides may be reflected through better reproduction accuracy when both straight sides are extended from their corners with the concave side and when the thick side is extended from its corner with the convex side.

Having established that reproduction accuracy reveals the intrinsic axial system of the Referent, it can now be considered if regional errors are still made with this Referent and axial system. Constructing an appropriate measure for regional errors with this Referent demands a somewhat more complicated procedure compared to experiment 1. Figure 12.9 illustrates the procedure globally; many assumptions are left implicit however.

The result of constructing the regional error like this is shown in Table 2. The distribution of the data in combination with the procedure to estimate regional error with four equal-sized areas has as a result that no strong statistical test is possible. It will, however, be assumed here that the differences between all categories are real.[8] As Table 12.2 shows, most regional errors are errors between the concave and convex regions that the thin–thick axis separates, errors between the thin and thick regions occur less often, and errors between both dimensions simultaneously hardly ever happen.

Of the two possible axial systems mentioned in van der Zee and Eshuis (Ch. 11), clearly only one has been used. This can be seen in the accuracy of reproduction, participants' reports, and in the regional errors that are made. The bounding box axes have not been found. Van der Zee and Eshuis's findings in this respect may be due to the task, which forced participants to find the best possible realization of directional prepositions: the findings may reflect the

TABLE 12.2 *Regional errors in experiment 2*

	Regional category				
	Correct	Concave/ convex error	Thin/thick error	Concave/convex and thin/thick error	TOTAL
No. of cases	6,191	202	64	9	6,466
(%)	95.7	3.1	1,0	0.1	100

[8] If each of the scores within each of the equal-sized ellipsoid areas were treated as if they were independent (which they are not), the differences among all four regional categories would be significant.

spatial opposition of the asked-for pair of opposing directional prepositions on the one hand, and the orthogonal orientation of the two pairs of opposing directional prepositions on the other. **Cuboids** (or **bounding boxes**) are then the ideal reference objects for directional prepositions (reflect the preferred arrangement of sides and axes). The categorical strategies mentioned by the participants in the present experiment do not show any consistent use of directional prepositions. The division of space in regions seems to be more flexible and participant's regional category descriptions ad hoc. Interestingly, participants still seem to be using opposing pairs and corresponding regional errors have been found. That more errors are made between 'convex' and 'concave' regions than between 'thick' and 'thin' regions corresponds nicely to the findings of van der Zee and Eshuis. Directional marking of the thin–thick axis appears to be the primary spatial division of this object: This is reflected in the relative amount of regional errors found here as well as in the preferred categorization of the axis as the front–back axis in van der Zee and Eshuis. The findings of van der Zee and Eshuis relating to the preferred categorization of axes seem to be transferable to non-linguistic spatial tasks.

3. Axes and Regions

One interesting observation that can be made regarding accuracy of reproduction is that relatively high accuracy is not restricted to axes, but can be seen as well (1) at locations close to axes, and—possibly—(2) at locations on imagined extensions of the Referent's contour. This is in line with recent findings of Munnich, Landau, and Dosher (2001; see also Landau, Ch. 2) where the findings of Hayward and Tarr (1995) are replicated with respect to linguistic and non-linguistic data for locations on axes and locations that are relatively far away from axes. Differences, however, appeared between the kinds of data for locations closer to the axes. While the linguistic task produces strong axial effects, the non-linguistic task (identity judgement) produces more graded effects with no apparent difference between locations on axes and locations that are close to the axes but still within the area between extensions of the Referent's contour. Whether or not axes are the primary determinant of local accuracy of spatial reproduction (or of identity judgement performance) is one issue. Another issue is what such axial effects have to do with spatial language. For one thing, the axial effects in non-linguistic spatial tasks thus far do not show any effect of half-axes whereas the linguistic spatial tasks do (see also Landau, Ch. 2). The experiments reported in this chapter do show directionality effects in that regional errors seem to occur from one half-plane to the complementary half-plane (half-planes separated by an axial boundary). This suggests that axes which function as boundaries may be underlying reproduction accuracy and that the categorization of regions (and directionality assigned to half-axes) may be responsible for the amount of regional error in the non-linguistic spatial task reported here.

One may argue that with two axes, four mutually exclusive regions are constructed where the half-axes function as boundaries (as in Crawford, Regier, and Huttenlocher, 2000). In that case one would expect that regional errors should occur more or less randomly over these four quadrants. This, however, does not appear to be the case. On the one hand, regional errors are typically made from a half-plane at one side of an axis to the half-plane at the other side of the axis with seemingly intact fine-grained information (these errors are thus between neighboring quadrants). Errors from one quadrant to the opposing (non-neighboring) quadrant hardly ever happen. Moreover, it can be shown that the amount of errors made from one half-plane to the opposing one is different from the amount of errors between the orthogonally oriented pair of half-planes (especially in experiment 1). This signals that the individual axes are the main organizing entities carving up space into regions and that the two axes (and the dimensions they separate) might be relatively independent.

Although axes do function as boundaries, this is only a part of the story. A front–back axis does not function only as the boundary between left and right, its half-axes also function as the best examples of front and back, just because the axis is neutral with respect to left and right. Thus, although axes, and perhaps an enhanced acuity for directional information associated with the axes, may be at the basis of the categorization of space, their prime function at the spatial level may be to function as boundaries between half-planes. Linguistically, one may refer to a direction unambiguously when this direction coincides with a half-axis. The spatial representation of intrinsic directions is considered as a combination of space categories (regional oppositions) and vector information. Activation of all spatial oppositions allows the vector to be oriented and shows which directions are the best examples of a given spatial dimension.

The goal of presenting an account of intrinsic object-centered frames of reference as has been done in this chapter is to make it plausible that object axes may function as boundaries in certain spatial tasks. The presented view is limited in scope though. For one thing, the present view is restricted to intrinsic object-centered frames of reference in the 2-D case although generalization to 3-D is possible. Regions surrounding Referents are associated with differences between object parts. This account may not necessarily be transferable to other frames of reference. For instance, egocentric frames of reference (intrinsic or relative) may be dependent on a line of sight, which may function as a basis of defining directions that does not require any boundaries or object parts. This may also be the case with environment-centered frames of reference in which the gravitational vertical plays a role. The relative merits of 'axis-as-prototype' accounts and 'axis-as-boundary' accounts is an empirical question. Mixtures of these approaches also seem to be possible. For instance, a front–back axis and an above–below axis may be defined as prototypes while at the same time the plane containing both axes may function as the boundary between left and right regions.

13

Spatial Prepositions, Spatial Templates, and 'Semantic' versus 'Pragmatic' Visual Representations

KENNY COVENTRY

Abstract

This chapter argues that an account of the use and comprehension of spatial prepositions is likely to involve at least two dimensions, one geometric and one extra-geometric. In the first section the nature of frame of reference instantiation with reference to spatial template theory (e.g. Logan and Sadler, 1996) is considered. The second section moves on to consider the results of a number of recent experiments employing the prepositions *over*, *under*, *above*, and *below*, and the implications the findings have for spatial template theory. In the final section the net is widened to consider a range of prepositions, including those that have been classified variously as 'topological', 'projective', and 'proximity' or 'directional'. It is argued that the types of representation needed to serve spatial language are unlikely to be of a single format. Furthermore it is argued that the types of representation generated from the visual system underlying spatial language use and comprehension are likely to involve both 'semantic' and 'pragmatic' components (Jeannerod, 1994).

1. Spatial Templates Underlying Spatial Language and Spatial Relations

Spatial prepositions cover a range of different types of terms that are unlikely to be subject to exactly the same constraints. For example, *across* clearly implies a path that can be captured (at least partially) using simple **vectors**, *in front of* involves the (potential) use of multiple **frames of reference**, while the use and comprehension of *at* is dependent on both the relative distance between objects and whether the objects are interacting (e.g. *the man is at the piano*).

The idea that there may be a common spatial representation system underlying the apprehension of both spatial language and spatial relations has been gaining popularity recently (see e.g. Bryant, 1997). The basis for this claim resides largely in a series of empirical demonstrations that show that judgements of spatial relations and judgements of the appropriateness of spatial prepositions to describe spatial relations are highly correlated. The focus in this section will be to examine the notion that the use of spatial templates underlies the apprehension of spatial relations and spatial language (a framework nicely summarized by Logan and Sadler, 1996; see also Hayward and Tarr, 1995). Logan and Sadler (1996) offer an unusually well-specified account of how geometric regions operate that is compatible with less well-specified linguistic accounts (e.g. Bennett, 1975; Herskovits, 1986).

Logan and Sadler claim that **spatial templates** underlie the apprehension of spatial relations and spatial prepositions. In order to work out whether a spatial relation applies in a given situation, it is proposed that a spatial template (or templates) representing regions of acceptability for the relevant relation is (or are) mapped onto a visual scene. The template is a representation that is centered on the reference object and is aligned with the reference frame imposed on or extracted from the reference object. In this way spatial template theory is an axis-based approach to specifying goodness of fit for relative positions of objects.

A useful way of thinking about spatial templates is to imagine a camera lens that can be imposed on the visual scene in a number of ways. On the viewfinder imagine there is a grid, which for the sake of argument, consists of 7 columns and 7 rows, each spaced equally apart (see Fig. 13.1). Therefore, there are 49 different points on the grid, and each point can be represented by a two point value (x, y), where x is a coordinate point on the horizontal axis and y is a point on the vertical axis. Imagine further that the lens has been zoomed in on the visual scene such that the reference object (or Ground, in this case a *square*) is centered on the grid at the midpoint (points (4,4)). For each set of coordinates on the grid there is an acceptability weighting, and the weightings on the template can then be used to assess the appropriateness of a given relation with respect to the scene the lens has zoomed in on. In this case for the term *above*, the letters marked on the grid represent good, acceptable, and bad regions for that term (denoted by G, A, and B respectively—following Carlson-Radvansky and Logan, 1997).

For Logan and Sadler (1996) different spatial relations are associated with different spatial templates, each containing graded ratings across their coordi-

A	A	A	G	A	A	A
A	A	A	G	A	A	A
A	A	A	G	A	A	A
B	B	B	■	B	B	B
B	B	B	B	B	B	B
B	B	B	B	B	B	B
B	B	B	B	B	B	B

FIG. 13.1 Acceptability ratings for the spatial relation *above*, according to Logan and Sadler (1996), as modified by Carlson-Radvansky and Logan (1997).

nates. Computing goodness of fit is a relatively straightforward matter, and is not the concern of this chapter. What is more important is the notion of how a spatial template is established, and how spatial templates map onto the myriad types of relation associated with individual lexical items.

In order to map a spatial template onto a visual scene it is necessary in some way to bind the arguments of the relation in the stored spatial template to the objects in the perceptual representation. In other words, what Logan and Sadler call *spatial indexing* is required to establish the correspondence between a symbol and a percept. Once spatial indexing has taken place, reference frame adjustment is required in order to extract the relevant reference frame from the reference object, to choose a scale, and a direction as appropriate. Going back to our camera lens analogy, the camera needs to be moved around so that the picture is the right way up, and the camera needs to be zoomed in as appropriate in order to achieve the right scale.

We can see how this kind of approach operates by considering consistent findings across a range of experimental paradigms. For example, Logan, and Sadler report the results of four experiments, which, they argue, support the spatial template framework. In a production task, participants were presented with a visual scene with the reference object (of a similar type to the square in Fig. 13.1) centered at the midpoint of the scene, and participants were asked to mark points on the picture which are appropriate for different spatial prepositions. The results produced consistent patterns across participants that were replicated in a rating paradigm in which participants had to rate the appropriateness of sentences involving spatial prepositions to describe each of the 49 coordinate points on a grid imposed over the visual scene. The results of these studies show that the most appropriate region for *above*, for example, is directly higher than the reference object (i.e. the regions denoted 'G' in Fig. 13.1). When the **Figure object** was positioned at the same point on the y-axis, but moved left or right of the center point of the reference objects on the x-axis, then the ratings went down. Similarly, moving the Figure object downwards on the y-axis was found to reduce appropriateness ratings. Therefore, the points marked in the production study map onto the highest ratings given in the rating study. A further experiment involved participants rating the similarity in meaning of word pairs using a multidimensional scaling approach. The results of this study again correlated with the results from the rating study just described. The similarity relations mapped onto the overlap in ratings for different terms and coordinates, and therefore Logan and Sadler argue that this provides evidence that people are indeed using spatial templates to compute relations.

Logan and Sadler contrast the spatial template view with the visual routine perspective presented by Ullman (1984). Ullman argued that perceptual processing requires specialized **visual routines** that operate on the output of basic low-level visual processing ('base' representations), and deliver a more flexible incremental interpretation of the visual scene. In relation to the experiments just reviewed, Logan and Sadler acknowledge that the results could be explained by visual routines, and set out to test whether this is the case. They suggest that Ullman's routines would predict a correlation between the time taken to make a judgement about an *above* relation and the distance between Figure and Ground, given that visual routines take time to apply in contrast to basic perceptual processes. In contrast, if multiple spatial templates are imposed on a scene in parallel, Logan and Sadler claim that the distance between Figure and Ground would not affect the time to make the judgement. In a reaction time study designed to test directly between these options, Logan and Sadler found no distance effects, and therefore they conclude that the spatial template view is supported. I will return to this finding later in the chapter.

One of the attractions of spatial template theory is that it offers a means of dealing with frame of reference ambiguity effects. One can use prepositions such as *in front of* or *above* with respect to one (or more) of three basic reference frames (Levinson, 1996). These are the **intrinsic** (or object-centered), **relative** (or viewer-deictic/deictic), or **absolute** (environment-deictic or extrinsic) frames (see also Introduction). The intrinsic frame locates a Figure with reference to the salient features of the Ground. For example, *the car is behind the house* used intrinsically locates the car in relation to the opposite wall from where the salient front of the house is, which is where the back door is. The relative use of the same expression would locate the car directly behind the opposite wall to the wall where the speaker and hearer are standing. The most common absolute use relates to the gravitational plane where such terms as *over* and *above* are used for positions higher than in the gravitational plane.

In relation to spatial template theory, it has been shown that multiple reference frames are active and indeed compete during spatial term assignment (e.g. Carlson-Radvansky and Irwin, 1994; Carlson-Radvansky and Logan, 1997). For example, Carlson-Radvansky and Irwin (1994), using a sentence-picture verification task, found that the use of a reference frame was slowed down when another reference frame indicated a conflicting direction for the same spatial term. These results provide important first steps in the understanding of spatial term assignment. However, as shall be demonstrated, the manipulation of reference frame conflicts does not influence all spatial terms in the same way.

2. Spatial Templates Reconsidered: The Case of *Over, Under, Above,* and *Below*

While the spatial template approach is appealing in that it offers a means of capturing goodness of fit relations for spatial relations and spatial language, there are a number of possible problems with the approach that can be considered in turn.

To begin with, it is assumed that there are different spatial templates for different spatial terms, and when polysemy exists, there is a different spatial template for each polyseme (Logan and Sadler, 1996: 499). This claim leads spatial template theory onto the same rocky ground encountered by fully specified accounts of lexical specification, as discussed in relation to prepositions by Coventry (1998). Take the case of *over*, for instance. Brugman (1988) and Lakoff (1987) have argued that this term is highly polysemic. While there are three central (prototypical) senses for *over* according to Brugman and Lakoff—the 'above' schema (where the Figure is higher than but not in contact with the Ground), the 'cover' schema (where the Figure covers the Ground and is usually above and in contact with the Ground as in *the table cloth is over the table*), and the 'above-across' schema (where the Figure is moving on a path above, and extending beyond the boundaries of the Ground as in *the plane flies over the bridge*)—there are dozens of less central senses all of which are labeled in their account. If Logan and Sadler are correct in their claim that there are unique spatial templates associated with each polyseme, then the use of multiple templates requires that some principles must be proposed to select which spatial template is appropriate in context. While this may be possible when there are only a few templates, it becomes computationally very expensive when the number of polysemes becomes large.

A second problem with spatial templates is that they don't appear to cope very well with a range of prepositions that involve the use of dynamics. For example, the 'above-across' schema for *over* (as in *the bird flew over the hill*) requires changes in regions of acceptability over time and is therefore unlikely to be captured in a single static spatial template. Templates in the form of the types of 'movies' proposed by Regier (1996) may be a more appropriate way of specifying path relations.

A third problem is that the use of spatial templates on their own does not account for a range of empirical findings that show that the positions of objects in space underdetermine the appropriateness of spatial terms. Other factors, which have to do with what objects are and how they interact with each other, have been shown to be central to the comprehension and production of spatial prepositions. Here we consider several cases (but see also Coventry, 1998 and Carlson, 2000).

Coventry, Prat-Sala, and Richards (2001) document findings which indicate that a number of modifications need to be made to spatial template theory. In Logan and Sadler's experiments, they found consistency in relation to the nature of spatial templates across a range of spatial terms. For example, for *over* and *above*, regions with the highest ratings are depicted in Fig. 13.1. However, as Coventry, Prat-Sala, and Richards note, Logan and Sadler used abstract objects in their studies, and this may have led to an artificial exaggeration of the types of effect they have produced (a point also made by Carlson-Radvansky, Covey, and Lattanzi, 1999). Coventry, Prat-Sala, and Richards presented

FIG. 13.2 Contrary to predictions by spatial template theory acceptability ratings for *over* and *above* are related to the degree of rotation and the functionality of the Figure (from Coventry, Prat-Sala, and Richards, 2001).

Spatial Prepositions   261

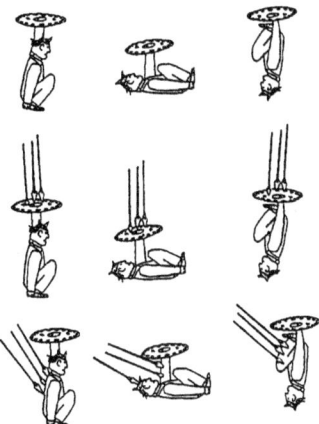

FIG. 13.3 Figure and Ground object relations aligned in the same way in different reference frames (column 1), or not aligned in different reference frames (columns 2 and 3).

participants with scenes of the type depicted in Figs. 13.2 and 13.3. Rather than using abstract objects, they used objects with functions (in the case of Fig. 13.2, protecting functions). Participants had the task of rating the appropriateness of sentences of the form *the Figure is preposition the Ground* in order to describe positions of objects in each scene. The prepositions used were *over*, *under*, *above*, and *below*. According to the spatial template view, the highest ratings for *above* (for example) should be directly higher than the ground (in line with the center of mass of the ground; see Regier, 1996).

Across a range of materials and experiments, when a single frame of reference is involved (or where frames of reference coincide), indeed the ratings given overall were related to the degree of rotation of the Figure from the vertical plane. Additionally it was found that **functional relations** are important determinants of the rating of the prepositions. Overall ratings for functional scenes (Fig. 13.2, middle row) were higher than ratings for control scenes (Fig. 13.2, the top row), which in turn were higher than ratings for the non-functional scenes (Fig. 13.2, bottom row). On their own, these findings could be regarded as a demonstration that spatial templates can be modified using **extra-geometric information** present in the visual scene being described. This suggests that more than geometric constraints are necessary for an account of the use and comprehension of spatial terms, a view that is now widely recognized (e.g. Aurnague, 1995; Carlson-Radvansky, Covey and Lattanzi, 1999; Coventry, 1998; Landau and Munnich, 1998; Garrod, Ferrier, and Campbell, 1999; Talmy, 1988; Vandeloise, 1994).

However, two further findings indicate greater problems for spatial template theory. First, the functional manipulation was found to influence the ratings of prepositions even in cases where the optimal place on the spatial template maps onto the scene. Even when the Figure was positioned directly above the Ground, non-functional scenes were given significantly lower ratings than functional and control scenes. If it is the case that functional relations come into play only when the prototypical geometric relation does not hold (as has been suggested by Landau and Munnich, 1998), then one might expect functional relations to have an effect on ratings only when the Figure is rotated away from the vertical plane. The fact that functional relations influence ratings across all positions suggests that multiple constraints are used to evaluate the appropriateness with which spatial expressions map onto visual scenes. This result is consistent with some recent results published by Carlson-Radvansky, Covey, and Lattanzi (1999), who found that regions of acceptability on a spatial template related to the center of mass of a Ground object could be modified by the introduction of a functional relation between the Figure and the Ground.

While there is evidence to suggest that geometric and extra-geometric relations are both important determinants of spatial language comprehension, Coventry, Prat-Sala, and Richards also provide evidence that different prep-

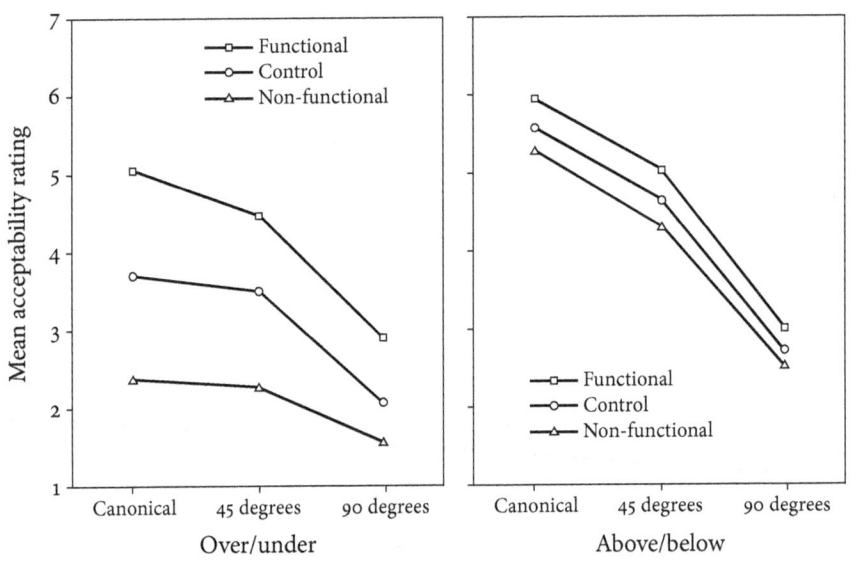

FIG. 13.4 Interaction between geometry, function, and preposition set found in Coventry, Prat-Sala, and Richards (2001) when frames of reference coincide.

ositions are affected by geometric and extra-geometric manipulations to varying degrees. For example, Fig. 13.4 illustrates a three-way interaction between function, geometry, and preposition set (*over/under* versus *above/below*) found in one of their experiments.

The geometric manipulation in this case was the position of the located object either canonically, rotated 45 degrees or rotated 90 degrees (as depicted in Fig. 13.2). As is clearly illustrated, *over* and *under* were more affected by functional relations than *above* and *below*, and *above* and *below* were more affected by the geometry manipulation than *over* and *under*. This is the first evidence to suggest that prepositions are influenced differentially by function and geometry, and in particular highlights that ratings predicted by spatial template theory may be more directly applicable to *above/below* than to *over/under*.

More striking differences between preposition sets are reported by Coventry, Prat-Sala, and Richards in an experiment that introduced conflicts between reference frames. Rather than rotating the Figure (as in Fig. 13.2), another experiment rotated the position of the Ground, therefore introducing conflicts between reference frames. Consider the scenes depicted in Fig. 13.3. With scenes on the left, the extrinsic (gravitational) and intrinsic (object-centered) frames of reference coincide. With scenes in the middle and on the right the extrinsic and intrinsic frames do not coincide, but conflict. For scenes on the left, one can say that *the shield is above the Viking* for both frames of reference, but for scenes in the middle and on the right *the shield is above the Viking* is appropriate for the extrinsic frame of reference but inappropriate for the intrinsic frame of reference (as the shield is not positioned higher than the Viking's head). Given that *over/under* versus *above/below* were found to be differentially affected by geometry (and function) when frames of reference coincided in the experiment described above, it was predicted that frame of conflict effects (also a change in the relative positions of objects with respect to their intrinsic axes) would be greater for *above/below* than for *over/under*. Indeed, the pattern of results found little effect of reference frame conflict for *over/under* although there were large function effects. Conversely, function effects for *above/below* were minimal while large frame of reference conflict effects were present for these terms. These patterns are illustrated in Figs. 13.5 and 13.6.

As can be seen in Fig. 13.5, *above* and *below* exhibit frame of reference conflict effects similar to those reported previously (e.g. Carlson-Radvansky and Irwin, 1994; Carlson-Radvansky and Logan, 1997); the greater the conflict (rotation of Ground), the lower the ratings for these prepositions. However, the results for *over* and *under* are quite different. While there is no difference between ratings for the canonical and 180-degree scenes (see Fig. 13.3 for

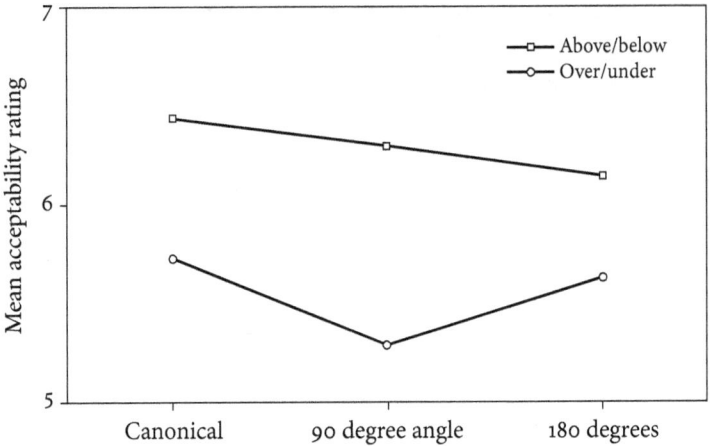

FIG. 13.5 Interaction between geometry and preposition set found in Coventry, Prat-Sala, and Richards (2001) when frames of reference conflict.

example scenes), when the Ground is rotated 90 degrees the ratings for *over/under* drop quite dramatically. Coventry, Prat-Sala, and Richards suggest that this result makes sense if one interprets it from a functional perspective. When the Ground is in the supine position, the surface area of the Figure is not large enough to ensure that the rain, for example, will not drip off the umbrella and

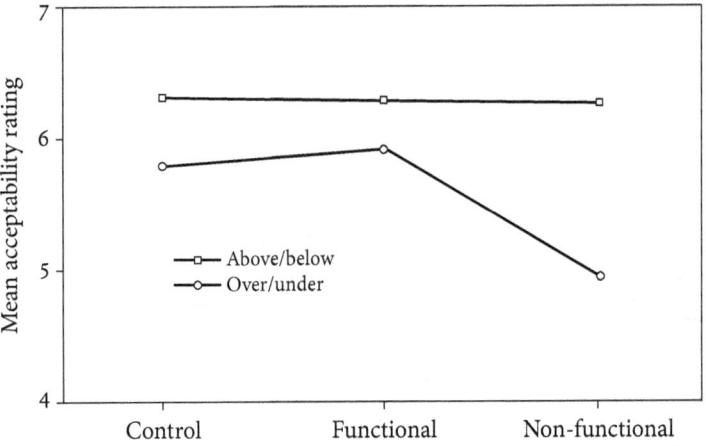

FIG. 13.6 Interaction between function and preposition set found in Coventry, Prat-Sala, and Richards (2001) when frames of reference conflict.

wet the man. Therefore the Figure is not able to fulfill its function due to its inappropriate size, and as a consequence the functional scenes and control scenes are much nearer the ratings for the non-functional scenes where the Figure is not fulfilling its function, although the non-functional ratings were still significantly lower than the other two levels of functionality. If this explanation is correct one might expect that the use of larger objects would cause this effect to disappear.

Overall, the results reported by Coventry, Prat-Sala, and Richards show the relative importance of geometry and extra-geometric variables for different prepositions. While Carlson-Radvansky and Radvansky (1996) have shown that functional relations can influence the choice of reference frame for a range of projective prepositions, the results reviewed go one stage further than this. While the results for *above/below* support Logan and Sadler's (1996) claim that multiple spatial templates may be imposed on the visual scene centered on the position of the Ground, the results for *over* and *under* suggest that there is no need to use multiple spatial templates for these terms as early processing of information in the scene (e.g. the rain) clearly primes the construction or selection of the absolute (environment-centered) frame. Therefore Logan and Sadler's (1996) claim that spatial templates are computed in parallel may not hold for *over/under*.

3. Casting a Wider Net: Spatial Prepositions and Spatial versus Action-Based Representations

The results reviewed for *over*, *under*, *above*, and *below* show that the comprehension of spatial prepositions involves multiple sources of information that are not weighted consistently across different prepositions. Here I wish to broaden out the discussion to consider a wider range of prepositions, and to address the issue of the types of spatial representation that are likely to serve spatial language.

While a geometric analysis of the type proposed by Logan and Sadler (1996: 526) offers a means of capturing geometric relations in a precise way, Logan and Sadler themselves indicate that the spatial template account is unlikely to 'capture the meanings of all spatial relations'. Furthermore, they give the example that the notions of containment and support associated with *in* and *on* imply more than just spatial relations. Indeed, a number of researchers have shown that the use and comprehension of *in* and *on* is influenced by what has been termed **'functional' or 'locational' control** (Garrod and Sanford, 1989; Coventry, 1998; Garrod, Ferrier, and Campbell, 1999; Vandeloise, 1994), a concept related to the force dynamics outlined by Talmy (1988). Furthermore, it has been shown that a term such as *in* can be used in cases where the geometric constraint of enclosure does not hold (Garrod and Sanford, 1989; Coventry, 1999). These results indicate the importance of what Garrod, Ferrier, and Campbell (1999) have termed **'functional geometry'**, a view that is developed in Coventry and Garrod (forthcoming). What is important about the spatial world is not just where objects are located in relation to one other, but how objects interact with each other and how we interact with objects. In this sense spatial prepositions that have been primarily associated with the 'where' system as opposed to the 'what' system (cf. Landau and Jackendoff, 1993), can be regarded as denoting much richer information involving a combination of representations from both the 'what' and 'where' systems. Jeannerod (1994) has argued for a distinction between what he calls 'semantic' and 'pragmatic' visual representations underlying visual and motor imagery, which map onto the distinct neuronal pathways of the what and where systems (see also Goodale *et al.* 1991). Whereas semantic representations encode detailed information about objects in a scene, pragmatic representations encode visual properties in relation to an object's affordances. However, the perceptual properties that together comprise an object's affordances are not bound together in a single representation (as is likely to be the case for recognition), but rather each component of prehension is directly driven by the relevant object properties. There is indeed evidence that viewing an object results in a direct potentiation of the actions it affords (see e.g. Tucker and Ellis, 1998). Similarly, talking about the spatial world may indeed involve the instantiation of affordances possessed by objects. The results of Coventry, Prat-Sala, and Richards (2001) can be interpreted within this framework. An umbrella affords a protecting function, and therefore this affordance has an effect on the appropriateness of various prepositions to describe its location. Similarly, containers afford the function of constraining the location of objects over time, hence dynamic information regarding the likely fulfillment of function over time is important in the use and comprehension of *in* (Coventry, 1998: Garrod, Ferrier, and Campbell, 1999).

While both the pragmatic and semantic systems may be relevant for spatial prepositions, it has also been shown that prepositions appear to be differentially affected by these types of information. Differences between *over/under* and *above/below* perhaps suggest that multiple sources of information generated both from the lexicon and from the scene itself may be integrated on-line in a mental model of the type proposed by Garrod and Sanford (1989) and Coventry (1998), the output of which leads to a situation-specific use/comprehension of a spatial expression. The view here is in line with a growing literature that demonstrates that multiple constraints, both linguistic and non-linguistic, influence comprehension on line (e.g. Sanford and Garrod, 1998; Sedivy et al., 1999).

As Coventry, Prat-Sala, and Richards (2001) and Carlson-Radvansky, Covey, and Lattanzi (1999) have noted, it is unsurprising that geometry effects have been so strong in the past across a range of experiments involving abstract objects of the type used by Logan and Sadler (1996). Where abstract objects are used, geometric considerations are bound to dominate, as there are no recognizable objects depicted with obvious affordances. Perhaps the lack of support for Ullman's visual routines noted in the discussion of Logan and Sadler's experiments maybe a result of the lack of use of real objects. The notion of the affordance of an umbrella, for example, is consistent with the idea that a visual inspection is carried out in order to evaluate the orientation and position of rain, for example, prior to giving a judgement about the suitability of a spatial term. In this sense models rooted in visual attention are desirable (see Carlson, Regier, and Covey, Ch. 6), although judgements of appropriateness have to go beyond processing of the relative positions of objects alone.

At the same time, it is possible to create a context in which functional relations dominate (e.g. Coventry, 1999; Garrod and Sanford, 1989). Therefore, contrary to the commonly held view (nicely summarized by Bryant, 1997) that there is a single system underlying spatial language and spatial representation that is geometric in nature, the evidence points to the need for the instantiation of multiple systems of representation that are integrated 'on line'. Exactly how these multiple sources of information are integrated on line remains an exciting and important area for future investigation.

References

ALLEN, J. (1984). Towards a general theory of action and time. *Artificial Intelligence*, 23, 123–54.

ANDERSON, J. M. (1971). *The grammar of case: Towards a localistic theory*. Cambridge: Cambridge University Press.

ASHER, N., and SABLAYROLLES, P. (1995). A typology and discourse semantics for motion verbs and spatial PPs in French. *Journal of Semantics*, 12(1), 163–209.

——, and VIEU, L. (1995). Towards a geometry of common sense: A semantics and a complete axiomatisation of mereotopology, in C. Mellish (ed.), *Proceedings of the 14th International Joint Conference on Artificial Intelligence (IJCAI95)*. Montreal: Morgan Kaufmann, 846–52.

AURNAGUE, M. (1991). *Contribution à l'étude de la sémantique formelle de l'espace et du raisonnement spatial: la localisation interne en français: sémantique et structures inférentielles*. Ph.D. thesis, Université Paul Sabatier.

——(1995). Orientation in French spatial prepositions: Formal representations and influences. *Journal of Semantics*, 12, 239–67.

——, and VIEU, L. (1993). A three-level approach to the semantics of space, in C. Zelinski-Wibbelt (ed.), *Semantics of prepositions in natural language processing*. Berlin: Mouton de Gruyter, iii. 393–439.

——, BORILLO A., BORILLO M. and VIEU L. (1991). *Sémantique de l'espace et cognition*. Paper presented at the International Cognitive Science Conference in San Sebastian.

BEHRMANN, M. (1999). Spatial reference frames and hemispatial neglect, in M. S. Gazzaniga (ed.), *The new cognitive neurosciences*. 2nd edn. Cambridge, Mass.: MIT Press, 651–66.

BELLUGI, U., MARKS, S., BIHRLE, A., and SABO, H. (1988). Dissociation between language and cognitive functions in Williams Syndrome, in D. Bishop and K. Mogford (eds.), *Language development in exceptional circumstances*. Hillsdale, NJ: Lawrence Erlbaum, 177–89.

——, BIHRLE, A., NEVILLE, H., DOHERTY, S., and JERNIGAN, T. L. (1992). Language, cognition, and brain organization in a neurodevelopmental disorder, in M. Gunnar and C. Nelson (eds.), *Developmental behavioral neuroscience: The Minnesota symposia on child psychology*. Hillsdale, NJ: Lawrence Erlbaum.

BENNETT, D. C. (1975). *Spatial and temporal uses of English prepositions. An essay in stratificational semantics*. London: Longman.

BIEDERMAN, I. (1987). Recognition by components: A theory of human image understanding. *Psychological Review*, 94, 115–47.

BIERWISCH, M. (1967). Some semantic universals of German adjectivals. *Foundations of Language*, 3, 1–36.

——(1996). How much space gets into language?, in P. Bloom, M. A. Peterson, L. Nadel, and M. F. Garrett (eds.), *Language and Space*. Cambridge, Mass.: MIT Press, Bradford Books, 31–76.

——, and LANG, E. (1989). *Dimensional Adjectives*. Berlin: Springer-Verlag.

BISIACH, E. (1996). Unilateral neglect and the structure of space representation. *Current Directions in Psychological Science*, 5(2), 62–5.

BLOOM, P., PETERSON, M. A.. NADEL, L., and GARRETT, M. F. (1996). *Language and Space*. Cambridge, Mass.: MIT Press, Bradford Books.

BOHNEMEYER, J. (1997). Yucatec Mayan lexicalization patterns in time and space, in M. Biemans and J. van der Weijer (eds.), *Proceedings of the CLS Opening Academic Year 1997-1998*. Tilburg: Center for Language Studies, 73–106.

——(submitted). The pitfalls of getting from here to there: Bootstrapping the syntax and semantics of motion event expressions in Yucatec Maya, in M. Bowerman and P. Brown (eds.), *Cross-linguistic perspectives on argument structure: Implications for learnability*. Proceedings of a workshop held at Max Planck Institute for Psycholinguistics, Nijmegen, June 1998.

——, and STOLZ, C. (submitted). The expression of spatial reference in Yukatek Maya: A survey. Submitted to S. C. Levinson and D. P. Wilkins (eds.), *The Grammar of Space*.

BORGO, S., GUARINO, N., and MASOLO, C. (1996). A pointless theory of space based on strong connection and congruence, in L. Carlucci Aiello and S. Shapiro (eds.), *Proceedings of principles of knowledge representation and reasoning (KR'96)*. Montreal: Morgan Kaufmann, 220–9.

BRUGMAN, C. (1988). *The story of 'over': Polysemy, semantics and the structure of the lexicon*. New York: Garland Press.

BRYANT, D. J. (1997). Representing space in language and perception. *Mind and Language*, 12(3/4), 239–64.

——, and TVERSKY, B. (1999). Mental representations of spatial relations from diagrams and models. *Journal of Experimental Psychology: Learning, Memory and Cognition*, 25, 137–56.

——, —— and FRANKLIN, N. (1992). Internal and external spatial frameworks for representing described scenes. *Journal of Memory and Language*, 31, 74–98.

——, —— and LANCA, M. (2000). Retrieving spatial relations from observation and memory, in E. van der Zee and U. Nikanne (eds.), *Cognitive interfaces: Constraints on linking cognitive information*. Oxford: Oxford University Press, 116–39.

CARLSON, L. (1999). Selecting a reference frame. *Spatial Computation and Cognition*, 1, 365–79.

——(2000). Object use and object location: The effect of function on spatial relations, in E. van der Zee and U. Nikanne (eds.), *Cognitive interfaces: Constraints on cognitive information*. Oxford: Oxford University Press, 94–115.

CARLSON-RADVANSKY, L. A., and IRWIN, D. E. (1993). Frames of reference in vision and language: Where is above? *Cognition*, 46, 223–44.

——(1994). Reference frame activation during spatial term assignment. *Journal of Memory and Language*, 33, 646–71.

——, and JIANG, Y. (1998). Inhibition accompanies reference-frame selection. *Psychological Science*, 9(5), 386–91.

——, and LOGAN, G. D. (1997). The influence of reference frame selection on spatial template construction. *Journal of Memory and Language*, 37, 411–37.

——, and RADVANSKY, G. A. (1996). The influence of functional relations on spatial term selection. *Psychological Science*, 7, 56–60.

——, COVEY, E. S., and LATTANZI, K. M. (1999). 'What' effects on 'where': Functional influences on spatial relations. *Psychological Science*, 10, 516–21.

CIENKI, A. (1998). STRAIGHT: An image schema and its metaphorical extensions. *Cognitive Linguistics*, 9(2), 147–9.

CLARK, E. V. (1973). Non-linguistic strategies and the acquisition of word meaning. *Cognition*, 2(2), 161–82.

——(1980). Here's the top: Nonlinguistic strategies in the acquisition of orientational terms. *Child Development*, 51(2), 329–38.

CLARK, H. H. (1973). Space, time, semantics and the child, in T. E. Moore (ed.), *Cognitive development and the acquisition of language*. New York: Academic Press, 27–64.

——(1996). *Using language*. Cambridge: Cambridge University Press.

CLARKE, B. (1985). Individuals and points. *Notre Dame Journal of Formal Logic*, 26(1), 61–75.

COLBY, C. L., and GOLDBERG, M. E. (1999). Space and attention in parietal cortex. *Annual Review of Neuroscience*, 22, 319–49.

COOK, W. A. (1989). *Case grammar theory*. Washington DC: Georgetown University Press.

CORBALLIS, M. C., and CULLEN, S. (1986). Decisions about the axes of disoriented shapes. *Memory and Cognition*, 14(1), 27–38.

COVENTRY, K. R. (1998). Spatial prepositions, functional relations, and lexical specifications, in Patrick Olivier and Klaus-Peter Gapp (eds.), *Representation and processing of spatial relations*. Mahwah, NJ: Lawrence Erlbaum, 247–62.

——(1999). Function, geometry and spatial prepositions: Three experiments. *Spatial Cognition and Computation*, 2, 145–54.

——, and GARROD, S. C. (forthcoming). *Saying, seeing and acting. The psychological semantics of spatial prepositions*. Brighton: Taylor Francis Psychology Press. Essays in Cognitive Psychology Series.

——, and PRAT-SALA, M. (1998). Geometry, function and the comprehension of 'over', 'under', 'above', and 'below', in M. A. Gernsbacher and S. J. Derry (eds.), *Proceedings of the Twentieth Annual Conference of the Cognitive Science Society*. Mahwah, NJ: Lawrence Erlbaum, 261–6.

—— ——, and RICHARDS, L. (2001). The interplay between geometry and function in the comprehension of Over, Under, Above, and Below. *Journal of Memory and Language*, 44(3), 376–98.

CRAWFORD, L. E., REGIER, T., and HUTTENLOCHER, J. (2000). Linguistic and non-linguistic spatial categorization. *Cognition*, 75(3), 209–35.

DAHL, R. (1977). *The wonderful story of Henry Sugar and six more*. New York: Bantam Books.

DENIS, M. (1997). The description of routes: A cognitive approach to the production of spatial discourse. *Cahiers de Psychologie Cognitive*, 16, 409–58.

DICKINSON, S. J., BERGEVIN, R., BIEDERMAN, I., EKLUNDH, J., MUNCK-FAIRWOOD, R., JAIN, A. K., and PENTLAND, A. (1997). Panel report: The potiential of geons for generic 3-D object recognition. *Image and Vision Computing*, 15, 277–92.

DUGAT, V., GAMBAROTTO, P., LARVOR, Y. (1999). Qualitative Theory of Shape and Structure, in P. Barahona and J. J. Alferes (eds.), *Lecture notes in artificial intelligence, LNAI-1695*. Berlin: Springer-Verlag, 148–62.

EHRICH, V. (1996). Verbbedeutung und Verbgrammatik: Transportverben im Deutschen, in E. Lang and G. Zifonun (eds.), *Deutsch—typologisch*. Berlin: Mouton de Gruyter, 229–60.

ESCHENBACH, C. (1999). Geometric structures of frames of reference and natural language semantics. *Spatial Computation and Cognition*, 1, 329–48.

——, and KULIK, L. (1997). An axiomatic approach to the spatial relations underlying 'left'–'right' and in 'front of'–'behind', in G. Brewka, C. Habel, and B. Nebel (eds.), *KI-97—Advances in Artificial Intelligence*. Berlin: Springer-Verlag, 207–18.

——, HABEL, C., and KULIK, L. (1999). Representing simple trajectories as oriented curves, in A. N. Kumar and I. Russell (eds.), *FLAIRS-99, Proceedings of the 12th International Florida AI Research Society Conference*. Orlando, Fla., 431–6.

——— ——, and LESSMOELLMANN, A. (1997). The integration of complex spatial relations by integrating frames of reference. Workshop on 'Language and Space', Fourteenth National Conference on Artificial Intelligence, Providence, Rhode Island, 27–31 July.

——, TSCHANDER, L., HABEL, C., KULIK, L. (2000). Lexical specifications of paths, in W. Brauer, C. Freksa, C. Habel, and K. F. Wender (eds.), *Spatial cognition II*. Berlin: Springer-Verlag, 131–48.

ESHUIS, R. (1995). *Spatial reference of relational nouns: The categorization of sides*. Unpublished master thesis, Utrecht University, The Netherlands.

——(in progress). *The categorization of space surrounding objects*. Doctoral dissertation. University of Hamburg.

EYSENCK, M. W., and KEANE, M. T. (1995). *Cognitive psychology: A student's handbook*. 3rd edn. Hove: Psychology Press.

FILLMORE, C. J. (1968). The case for case, in E. Bach and R. T. Harms (eds.), *Universals of linguistic theory*. New York: Holt, Rinehart, & Winston, 1–90.

——(1971). *Santa Cruz lectures on deixis*. Bloomington, Ind.: Indiana University Linguistics Club.

——, and KAY, P. (1997). Berkeley Construction Grammar. http://www.icsi.berkeley.edu/~kay/bcg/ConGram.html.

FORBUS, K. (1983). Qualitative reasoning about space and motion, in D. Gentner and A. Stevens (eds.), *Mental models*. Hillsdale, NJ: Lawrence Erlbaum, 53–73.

FRANK, A. (1992). Qualitative spatial reasoning about distances and directions in geographic space. *Journal of Visual Languages and Computing*, 3, 343–71.

FRANKLIN, N., and TVERSKY, B. (1990). Searching imagined environments. *Journal of Experimental Psychology: General*, 110, 63–76.

——, HENKEL, L. A., and ZANGAS, T. (1995). Parsing surrounding space into regions. *Memory and Cognition*, 23, 397–407.

——, TVERSKY, B., and COON, V. (1992). Switching points of view in spatial mental models acquired from text. *Memory and Cognition*, 20, 507–18.

FREKSA, C. (1992). Using orientation information for qualitative spatial reasoning, in A. Frank, I. Campari, and U. Formentini (eds.), *Theories and methods of spatio-temporal reasoning in geographic space. Proceedings of the International Conference GIS—From Space to Territory*, LNCS. Berlin: Springer-Verlag, 162–78.

——, HABEL, C., and WENDER, K. F. (1998). *Spatial cognition: An interdisciplinary approach to representing and processing spatial knowledge*. Berlin: Springer-Verlag.

FRISK, V., and MILNER, B. (1990). The role of the left hippocampal region in the acquisition and retention of story content. *Neuropsychologia*, 28, 349–59.

FRIEDERICI, A. D., and LEVELT, W. J. M. (1990). Spatial reference in weightlessness: Perceptual factors and mental representations. *Perception and Psychophysics*, 47, 253–66.

GALLISTEL, C. R. (1990). *The organization of learning*. Cambridge, Mass.: MIT Press.

GALTON, A. (1997). Space, time and movement, in O. Stock (ed.), *Spatial and temporal reasoning*. Dordrecht: Kluwer, 321–52.

GAPP, K.-P. (1995). Angle, distance, shape, and their relationship to projective relations, in J. D. Moore and J. F. Lehman (eds.), *Proceedings of the 17th Annual Conference of the Cognitive Science Society*, Mahwah, NJ: Cognitive Science Society, 112–17.

GARNHAM, A. (1989). A unified theory of meaning of some spatial relational terms. *Cognition*, 31, 45–60.

GARROD, S. C., and SANFORD, A. J. (1989). Discourse models as interfaces between language and the spatial world. *Journal of Semantics*, 6, 147–60.

——, FERRIER, G., and CAMPBELL, S. (1999). 'In' and 'on': investigating the functional geometry of spatial prepositions. *Cognition*, 72, 167–89.

GEORGOPOULOS, A. P., SCHWARTZ, A. B., and KETTNER, R. E. (1986). Neuronal population coding of movement direction. *Science*, 223, 1416–19.

GERLA, G. (1994). Pointless geometries, in F. Buekenhout (ed.), *Handbook of incidence geometry*. Amsterdam: Elsevier, 1015–31.

GIBSON, J. J. (1950). *The perception of the visual world*. Boston: Houghton Mifflin.

GOLDBERG, A. (1991). It can't go down the chimney up: Paths and the English resultative. *Proceedings of the Seventeenth Annual Meeting of the Berkeley Linguistics Society*, 368–78.

——(1995). *Constructions*. Chicago: University of Chicago Press.

GOODALE, M. A., MILNER, A. D., JAKOBSON, L. S., and CAREY, D. P. (1991). A neurological distinction between perceiving objects and grasping them. *Nature*, 349(10), 154–6.

GRICE, H. P. (1975). Logic and conversation, in P. Cole and J. L. Morgan (eds.), *Speech acts*. New York: Academic Press, 41–58.

GROUNDSTROEM, A. (1988). *Finnische Kasusstudien*. Acta Universitas Umensis. Umeå Studies in Humanities 87. University of Umeå.

HABEL, C. (1989). Zwischen-Bericht, in C. Habel, M. Herweg, and K. Rehkaemper (eds.), *Raumkonzepte in Verstehensprozessen: Interdisziplinäre Beiträge zu Sprach und Raum*. Tübingen: Niemeyer, 37–69.

——(1999). Drehsinn und Reorientierung—Modus und Richtung beim Bewegungsverb *drehen*, in G. Rickheit (ed.), *Richtungen im Raum*. Wiesbaden: Deutscher Universitäts-Verlag, 101–28.

HARNAD, S. (1987). *Categorical perception: The groundwork of cognition*. New York: Cambridge University Press.

HARRIS, L. J., and STROMMEN, E. A. (1974). What is the 'front' of a simple geometric form? *Perception and Psychophysics*, 15(3), 571–80.

HARTLEY, T., BURGESS, N., LEVER, C., CACUCCI, F., and O'KEEFE, J. (2000). Modelling place fields in terms of the cortical inputs to the hippocampus. *Hippocampus*, 10, 369–79.

HAYES, P. (1985). The second naive physics manifesto, in R. C. Moore and J. H. (eds.), *Formal theories of the commonsense world*. Norwood: Ablex Publishing Corporation, 1–36.

HAYS, E. (1989). On defining motion verbs and spatial prepositions. Technical report, Universitat des Saarlandes.

HAYWARD, W. G., and TARR, M. J. (1995). Spatial language and spatial representation. *Cognition*, 55, 39–84.

HEINE, B., CLAUDI, U., and F. HÜNNEMEYER (1991). *Grammaticalization: A conceptual framework*. Chicago: University of Chicago Press.

HELMANTEL, M. (1998). Simplex adpositions and vector theory. *The Linguistic Review*, 15, 361–88.

HERSKOVITS, A. (1986). *Language and spatial cognition: An interdisciplinary study of prepositions in English*. Cambridge: Cambridge University Press.

——(1998). Schematization, in P. Olivier and Gapp, K.-P. (eds.), *Representation and processing of spatial expressions*. Mahwah, NJ: Lawrence Erlbaum, 149–62.

HOFFMAN, J., LANDAU, B., and PAGANI, J. (2002). Spatial breakdown in spatial construction:

Evidence from eye fixations in children with Williams syndrome. *Cognitive Psychology* (forthcoming).

HOTHERSALL, D. (1995). *History of Psychology*. 3rd edn. New York: McGraw Hill.

HUTTENLOCHER, J., and STRAUSS, S. (1968). Comprehension and a statement's relation to the situation it describes. *Journal of Verbal Learning and Verbal Behavior*, 7, 300–4.

——, HEDGES, L. V., and DUNCAN, S. (1991). Categories and particulars: Prototype effects in estimating spatial location. *Psychological Review*, 98(3), 352–76.

ITKONEN, E. (1999) [1995]. Review of Ray S. Jackendoff, *The Languages of the Mind*, in E. Itkonen: *Kielitieteen kääntöpuoli*. Publications in General Linguistics, 2. University of Turku. [Published earlier in *Word*, 3 (1995).]

JACKENDOFF, R. (1983). *Semantics and cognition*. Cambridge, Mass.: MIT Press.

——(1990). *Semantic structures*. Cambridge, Mass.: MIT Press.

——(1991). Parts and boundaries. *Cognition*, 41, 9–45.

——(1996a). The proper treatment of measuring out, telicity, and perhaps even quantification in English. *Natural Language and Linguistic Theory*, 14, 305–54.

——(1996b). The architecture of the linguistic-spatial interface, in P. Bloom, M. A. Peterson, L. Nadel, and M. F. Garrett (eds.), *Language and space*. Cambridge, Mass.: MIT Press, Bradford Books, 1–30.

——(1997). *The architecture of the language faculty*. Cambridge, Mass.: MIT Press.

——, and LANDAU, B. (1992). Spatial language and spatial cognition, in R. Jackendoff, *The languages of the mind*. Cambridge, Mass.: MIT Press, 99–124.

JEANNEROD, M. (1994). The representing brain: Neural correlates of motor intention and imagery. *Behavioral and Brain Sciences*, 17(2), 187–245.

JOHNSON-LAIRD, P. N. (1983). *Mental models*. Cambridge, Mass.: Harvard University Press.

KAUFMANN, I. (1995). *Konzeptuelle Grundlagen semantischer Dekompositionsstrukturen. Die Kombinatorik lokaler Verben und prädikativer Komplemente*. Tübingen: Niemeyer.

KRIFKA, M. (1992). Thematic relations as links between nominal reference and temporal constitution, in I. Sag and A. Szabolcsi (eds.), *Lexical matters*. Stanford, Calif.: CSLI Publications, 29–54.

——(1995). Telicity in movement, in P. Amsili, M. Borillo, and L. Vieu (eds.), *Time, space and movement: Meaning and knowledge in the sensible world*. Toulouse: Université Paul Sabatier, 63–75.

LAKOFF, G. (1987). *Women, fire, and dangerous things*. Chicago: University of Chicago Press.

LANDAU, B. (1999). *Space and language in Williams syndrome*. Paper presented at the Conference on Spatial Intelligence, University of Chicago.

——(2003). Axial representations in language and cognition: Evidence from children with Williams syndrome. In preparation.

——, and JACKENDOFF, R. (1993). 'What' and 'Where' in spatial language and spatial cognition. *Behavioral and Brain Sciences*, 16, 217–65.

——, and MUNNICH, E. (1998). The representation of space and spatial language: Challenges for cognitive science, in P. Olivier and K.-P. Gapp (eds.), *Representation and processing of spatial expressions*. Mahwah, NJ: Lawrence Erlbaum, 263–72.

LANG, E. (1990). Primary perceptual space and inherent proportion schema: Two interacting categorization grids underlying the conceptualization of spatial objects. *Journal of Semantics*, 7, 121–41.

——(1991). A two-level approach to projective prepositions, in G. Rauh (ed.), *Approaches to prepositions*. Tübingen: Gunter Narr Verlag, 127–67.

LANGACKER, R. (1987). *Foundations of cognitive grammar*, i. *Theoretical Prerequisites*. Stanford, Calif.: Stanford University Press.
LEECH, G. (1969). *Towards a semantic description of English*. London: Longman.
LEIBOWITZ, H. W., GUZY, L. T., PETERSON, E., and BLAKE, P. T. (1993). Quantitative perceptual estimates: Verbal versus nonverbal retrieval techniques. *Perception*, 22, 1051–60.
LESNIEWSKI, S. (1992). *Collected works*, ed. S. J. Surma, J. T. Srzednicki, J. D. Barnett. Dordrecht: Kluwer; Warsaw: Polish Scientific Publishers, vols. i, ii.
LEVELT, W. J. M. (1982). Cognitive styles in the use of spatial direction terms, in R. J. Jarvella and W. Klein (eds.), *Speech, place and action*. Chichester: Wiley, 251–68.
——(1984). Some perceptual limitations on talking about space, in A. J. van Doorn, W. A. van der Grind, and J. J. Koenderink (eds.), *Limits in perception*. Utrecht: VNU Science Press, 323–58.
——(1989). *Speaking: From intention to articulation*. Cambridge, Mass.: MIT Press.
——(1996). Perspective taking and ellipsis in spatial descriptions, in P. Bloom, M. A. Peterson, L. Nadel, and M. Garrett (eds.), *Language and space*. Cambridge, Mass.: MIT Press, 77–107.
LEVIN, B. (1993). *English verb classes and alternations. A preliminary investigation*. Chicago: University of Chicago Press.
LEVINSON, S. C. (1992a). *Language and cognition: The cognitive consequences of spatial description in Guugu Yimithirr*. Working paper no. 13, Cognitive Anthropology Research Group, Max Planck Institute for Psycholinguistics, Nijmegen.
——(1992b). *Vision, shape and linguistic description: Tzeltal body-part terminology and object description*. Working paper No. 12, Cognitive Anthropology Research Group, Max Planck Institute for Psycholinguistics, Nijmegen.
——(1996). Frames of reference and Molyneux's questions: Cross-linguistic evidence, in P. Bloom, M. A. Peterson, L. Nadel, and M. Garrett (eds.), *Language and space*. Cambridge, Mass.: MIT Press, 109–69.
——(2000). *Presumptive meanings: The theory of generalised conversational implicatures*. Cambridge, Mass.: MIT Press.
LEYTON, M. (1992). *Symmetry, causality, mind*. Cambridge, Mass.: MIT Press.
LIN, E. L., and MURPHY, G. L. (1997). Effects of background knowledge on object categorization and part detection. *Journal of Experimental Psychology: Human Perception and Performance*, 23, 1153–69.
LINK, G (1983). The logical analysis of plural and mass terms: A lattice-theoretic approach, in R. Bäuerle, C. Schwarze, and A. von Stechow (eds.), *Meaning, use and interpretation of language*. Berlin: De Gruyter, 302–23.
——(1998). *Algebraic semantics in language and philosophy*. CSLI Lecture Notes No. 74. Stanford, Calif.: CSLI Publications.
LOGAN, G. D. (1994). Spatial attention and the apprehension of spatial relations. *Journal of Experimental Psychology: Human Perception and Performance*, 20, 1015–36.
——(1995). Linguistic and conceptual control of visual spatial attention. *Cognitive Psychology*, 28, 103–74.
——, and COMPTON, B. (1996). Distance and distraction effects in the apprehension of spatial relations. *Journal of Experimental Psychology: Human Perception and Performance*, 22, 159–72.
——, and SADLER, D. D. (1996). A computational analysis of the apprehension of spatial relations, in P. Bloom, M. A. Peterson, L. Nadel, and M. F. Garrett (eds.), *Language and space*. Cambridge, Mass.: MIT Press, 493–529.

McCloskey, M., and Rapp, B. (2000). Attention-referenced visual representations: Evidence from impaired visual localization. *Journal of Experimental Psychology: Human Perception and Performance*, 26, 917–33.

———, Yantis, S., Rubin, G., Bacon, W. F., Dagnelie, G., Gordon, B., Aliminosa, D., Boatman, D. F., Badecker, W., Johnson, D. N., Rusa, R. J., and Palmer, E. (1995). A developmental deficit in localizing objects from vision. *Psychological Science*, 6(2), 112–17.

Maienborn, C. (1990). *Position und Bewegung: Zur Semantik lokaler Verben.* IWBS-Report No. 138. Stuttgart: IBM.

——(1994). Kompakte Strukturen: Direktionale Präpositionen und nicht-lokale Verben, in S. W. Felix, C. Habel, and G. Rickheit (eds.), *Kognitive Linguistik. Repräsentationen und Prozesse*. Opladen: Westdeutscher Verlag, 229–49.

Marr, D. (1982). *Vision.* New York: Freeman.

Mainwaring, S. D., Tversky, B., Ogishi, M., and Schiano, D. J. (forthcoming). Descriptions of simple spatial scenes in English and Japanese. *Spatial Cognition and Computation.*

Mervis, C. B., Morris, C. A., Bertrand, J., and Robinson, B. F. (1999). Williams syndrome: Findings from an integrated program of research, in H. Tager-Flusberg (ed.), *Neurodevelopmental disorders.* Cambridge, Mass.: MIT Press, 65–110.

Miller, G. A., and Johnson-Laird, P. N. (1976). *Language and perception.* Cambridge, Mass.: Harvard University Press.

Morris, C. A., Ewart, A. K., Sternes, K., Spallone, P., Stock, A. D., Leppert, M., and Keating, M. T. (1994). Williams syndrome: Elastin gene deletions. *American Journal of Human Genetics*, 55 (Suppl.): A89.

Morrow, D. G., and Clark, H. H. (1988). Interpreting words in spatial descriptions. *Language and Cognitive Processes*, 3, 275–91.

Muller, P. (1998). Space-time as a primitive for space and motion, in N. Guarino (ed.), *Proceedings of the International Conference on Formal Ontology in Information Systems (FOIS98).* Amsterdam: IOS Press, 63–76.

——, and Sarda, L. (1999). Représentation de la sémantique des verbes de déplacements transitifs du français. *Traitement Automatique des Langues, ATALA*, 39(2), 127–47.

Munnich, E., and Landau, B. (1997). *Universals of spatial representation in language and memory.* Poster presented at the 38th Annual Meeting of the Psychonomic Society, Philadelphia.

———, and Dosher, B. (2001). Spatial language and spatial representation: A crosslinguistic comparison. *Cognition*, 81(3), 171–208.

Nam, S. (1995). *The semantics of locative prepositional phrases in English.* Ph.D. thesis, UCLA.

Nikanne, U. (1990). *Zones and tiers: A study of thematic structure.* Unpublished doctoral thesis, Finnish Literature Society, Helsinki, Finland.

——(1993). On assigning semantic cases in Finnish, in A. Holmberg and U. Nikanne (eds.), *Case and other functional categories in Finnish syntax.* Studies in generative grammar, 39. Berlin: Mouton de Gruyter, 75–87.

——(1995). Action tier formation and argument linking. *Studia Linguistica*, 49, 1–32.

——(1997a). Locative case adjuncts in Finnish: Notes on syntactico-semantic interface. *Nordic Journal of Linguistics*, 20, 155–78.

——(1997b). Lexical conceptual structure and syntactic arguments. *SKY*, 81–118.

——(1998). The lexicon and conceptual structure, in T. Haukioja (ed.), *Papers from the 16th Scandinavian Conference of Linguistics.* Turku: Publication 60 of the Department of Finnish and General Linguistics of the University of Turku, 305–19.

NIKANNE, U. (2000a). Some restrictions in linguistic expressions of spatial movement, in E. van der Zee and U. Nikanne (eds.), *Cognitive interfaces: Constraints on linking cognitive information*. Oxford: Oxford University Press, 77–93.

——(2000b) Constructions in Conceptual Semantics, in J.-O. Östman (ed.), *Construction Grammar(s): Cognitive and Cross-Language Dimensions*. Amsterdam/Philadelphia: John Benjamins.

O'KEEFE, J. (1990). A computational theory of the hippocampal cognitive map, in O. P. Ottersen and J. Storm-Mathiesen (eds.), *Understanding the brain through the hippocampus*. Progress in Brain Research, 83. Amsterdam: Elsevier, 287–300.

——(1996). The spatial prepositions in English, vector grammar, and the cognitive map theory, in P. Bloom, M. A. Peterson, L. Nadel, and M. F. Garrett (eds.), *Language and space*. Cambridge, Mass.: MIT Press, Bradford Books, 277–316.

——, and BURGESS, N. (1996). Geometric determinants of the place fields of hippocampal neurons. *Nature*, 381, 425–8

——, and NADEL, L. (1978). *The hippocampus as a cognitive map*. Oxford: Clarendon Press.

OLIVIER, P., and GAPP, K.-P. (1998). *Representation and processing of spatial expressions*. Mahwah, NJ: Lawrence Erlbaum.

PENTTILÄ, A. (1957). *Suomen kielioppi*. Porvoo: Verner Söderström.

PERRIG, W., and KINTSCH, W. (1985). Propositional and situational representations of text. *Journal of Memory and Language*, 24, 503–18.

PETERSON, M. A., NADEL, L., BLOOM, P., and GARRETT, M. F. (1996). Space and language, in P. Bloom, M. A. Peterson, L. Nadel, and M. F. Garrett (eds.), *Language and space*. Cambridge, Mass.: MIT Press, 553–77.

PHILBECK, J. W., LOOMIS, J. M., and BEALL, A. C. (1997). Visually perceived location is an invariant in the control of action. *Perception and Psychophysics*, 59(4), 601–12.

RANDELL, D., CUI, Z., and COHN, A. (1992). *A spatial logic based on regions and connection (KR'92)*. San Mateo, Calif.: Morgan Kaufmann.

RAUH, G. (1991). *Approaches to prepositions*. Tübingen: Gunter Narr Verlag.

REGIER, T. (1995). A model of the human capacity for categorizing spatial relations. *Cognitive Linguistics*, 6(1), 63–88.

——(1996). *The human semantic potential: Spatial language and constrained connectionism*. Cambridge, Mass.: MIT Press.

——(1997). Constraints on the learning of spatial terms: A computational investigation, in R. Goldstone, P. Schyns, and D. Medin (eds.), *Psychology of Learning and Motivation: Mechanisms of Perceptual Learning*, 36. San Diego, Calif.: Academic Press, 171–217.

——, and CARLSON, L. (2001). Grounding spatial language in perception: An empirical and computational investigation. *Journal of Experimental Psychology: General*, 130(2), 273–98.

ROEPER, P. (1997). Region-based topology. *Journal of Philosophical Logic*, 2, 251–309.

SADALLA, E. K., BURROUGHS, W. J., and STAPLIN, L. J. (1980). Reference points in spatial cognition. *Journal of Experimental Psychology: Human Learning and Memory*, 6, 516–28.

SANFORD, A. J., and GARROD, S. C. (1998). The role of scenario mapping in text comprehension. *Discourse Processes*, 26, 159–90.

SCHIANO, D. J., and TVERSKY, B. (1992). Structure and strategy in encoding simplified graphs. *Memory and Cognition*, 20(1), 12–20.

SCHMIDTKE, H. R. (1999). *Ein formaler Ansatz für die relative Lokalisierung ausgedehnter Objekte in 2-dimensionalen Layouts*. Diploma thesis. Department for Informatics. University of Hamburg. Available via www.informatik.uni-hamburg.de/WSV/Axiomatik-deutsch.html.

——(forthcoming). The house is north of the river: Relative localization of extended objects, in *Spatial information theory. Proceedings of COSIT '01*. Berlin: Springer-Verlag.

SCHOBER, M. F. (1993). Spatial perspective-taking in conversation. *Cognition*, 47, 1–24.

——(1995). Speakers, addressees, and frames of reference: Whose effort is minimized in conversations about locations. *Discourse Processes*, 20, 219–47.

SCHULZE, R. (1991). Getting round to (a)round: Towards a description and analysis of a spatial predicate, in G. Rauh (ed.), *Approaches to prepositions*. Tübingen: Gunter Narr Verlag, 251–74.

SCHYNS, P. G., GOLDSTONE, R. L., and THIBAUT, J.-P. (1998). The development of features in object concepts. *Behavioral and Brain Sciences*, 21, 1–54.

SEDIVY, J. C., TANENHAUS, M. K., CHAMBERS, C. G., and CARLSON, G. N. (1999). Achieving incremental interpretation through contextual representation. *Cognition*, 71, 109–47.

SPERBER, D., and WILSON, D. (1986). *Relevance: Communication and cognition*. Oxford: Blackwell.

SVOROU, S. (1994). *The grammar of space*. Amsterdam/Philadelphia: John Benjamins.

TALMY, L. (1972). *Semantic structures in English and Atsugewi*. Doctoral dissertation, University of California.

——(1983). How language structures space, in H. L. Pick, Jr. and L. P. Acredolo (eds.), *Spatial orientation: Theory, research and application*. New York: Plenum, 225–82.

——(1985). Lexicalization patterns, in T. Shopen (ed.), *Language typology and syntactic description. iii. Grammatical categories and the lexicon*. Cambridge: Cambridge University Press, 57–149.

——(1988). Force dynamics in language and cognition. *Cognitive Science*, 12, 49–100.

——(1991). Path to realization: A typology of event conflation, in L. A. Sutton, C. Johnson, and R. Shields (eds.), *Proceedings of the Seventeenth Annual Meeting of the Berkeley Linguistics Society. General Session and Parasession on the Grammar of Event Structure*. Berkeley: Berkeley Linguistics Society.

——(1996). Fictive motion in language and 'ception', in P. Bloom, M. A. Peterson, L. Nadel, and M. F. Garrett (eds.), *Language and space*. Cambridge, Mass.: MIT Press, Bradford Books, 211–76.

——(2000). *Toward a cognitive semantics. i. Concept-structuring systems. 2. Typology and process in concept structuring*. Cambridge, Mass.: MIT Press.

TARSKI, A. (1969). What is elementary geometry?, in J. Hintikka, (ed.), *The Philosophy of Mathematics*. Oxford: Oxford University Press, 164–75.

TAUBE, J. S., MULLER, R. U., and RANCK, J. B. (1990). Head-direction cells recorded from the post-subiculum in freely moving rats. I. Description and quantitative analysis. *Journal of Neuroscience*, 10, 420–35.

TAYLOR, J. (1995). *Linguistic categorization: Prototypes in linguistic theory*. 2nd edn. Oxford: Clarendon Press.

TAYLOR, H. A., and TVERSKY, B. (1992*a*). Descriptions and depictions of environments. *Memory and Cognition*, 20, 483–96.

————(1992*b*). Spatial mental models derived from survey and route descriptions. *Journal of Memory and Language*, 31, 261–82.

————(1996). Perspective in spatial descriptions. *Journal of Memory and Language*, 35, 371–91.

TELLER, P. (1969). Some discussion and extension of Manfred Bierwisch's work on German adjectivals. *Foundations of Language*, 5, 185–217.

TIPPER, S. P., and BEHRMANN, M. (1996). Object-centered not scene-based visual neglect. *Journal of Experimental Psychology: Human Perception and Performance*, 22(5), 1261–78.

TOLMAN, E. C. (1948). Cognitive maps in rats and men. *Psychological Review*, 5, 189–208.

TUCKER, M., and ELLIS, R. (1998). On the relations between seen objects and components of action. *Journal of Experimental Psychology: Human Perception and Performance*, 24(3), 830–46.

TVERSKY, B. (1996). Spatial perspective in description, in P. Bloom, M. A. Peterson, L. Nadel, and M. Garrett (eds.), *Language and space*. Cambridge, Mass.: MIT Press, 463–91.

——(2000a). Levels and structure of cognitive mapping, in R. Kitchin and S. M. Freundschuh (eds.), *Cognitive mapping: Past, present and future*. London: Routledge.

——(2000b). Remembering space, in E. Tulving and F. I. M. Craik (eds.), *Handbook of Memory*. New York: Oxford University Press.

——, and LEE, P. U. (1998). How space structures language, in C. Freksa, C. Habel, and K. F. Wender (eds.), *Spatial cognition: An interdisciplinary approach to representation and processing of spatial knowledge*. Berlin: Springer-Verlag, 157–75.

————, and MAINWARING, S. (forthcoming). Why speakers mix perspectives. *Journal of Spatial Cognitive and Computation*.

————(1999). Pictorial and verbal tools for conveying routes, in C. Freksa and D. M. Mark (eds.), *Spatial information theory: Cognitive and computational foundations of geographic information science*. Berlin: Springer-Verlag, 51–64.

——, KIM, J., and COHEN, A. (1999). Mental models of spatial relations and transformations from language, in C. Habel and G. Rickheit (eds.), *Mental models in discourse processing and reasoning*. Amsterdam: North-Holland, 239–58.

——, and SCHIANO, D. J. (1989). Perceptual and conceptual factors in distortions in memory for graphs and maps. *Journal of Experimental Psychology: General*, 118(4), 387–98.

——, TAYLOR, H. A., and MAINWARING, S. (1997). Langage et perspective spatial (Spatial perspectives in language), in M. Denis (ed.), *Langage et cognition spatiale*. Paris: Masson, 25–49.

ULLMAN, S. (1984). Visual routines. *Cognition*, 18, 97–159.

UNGERLEIDER, L. G., and MISHKIN, M. (1982). Two cortical visual systems, in D. J. Ingle, M. A. Goodale, and R. J. W. Mansfield (eds.), *Analysis of visual behavior*. Cambridge, Mass.: MIT Press, 549–86.

VANDELOISE, C. (1991). *Spatial prepositions: A case study from French*, trans. Anna R. K. Bosch. Chicago: University of Chicago Press. (Original work published 1986.)

——(1994). Methodology and analyses of the preposition 'in'. *Cognitive Linguistics*, 5(2), 157–84.

VAN DER ZEE, E. (1996). *Spatial knowledge and spatial language: A theoretical and empirical investigation*. Utrecht: ISOR Publications.

——(1997). A case study in conceptual semantics: The conceptual structure of Dutch curl verbs, fold verbs and crumple verbs, in B. Kokinov (ed.), *Perspectives in Cognitive Science 4*. Sofia: NBU Press.

——(2000). Curvature representation in the lexical interface, in E. van der Zee and U. Nikanne (eds.), *Cognitive interfaces: Constraints on linking cognitive information*. Oxford: Oxford University Press, 143–82.

——, and ESHUIS, R. (1997). *Spatial language and spatial cognition: The categorization of sides and regions*, Proceedings of the Workshop in Spatial Cognition in Rome, Italy, 73–4.

——, and NIKANNE, U. (2000). *Cognitive interfaces: Constraints on linking cognitive information*. Oxford: Oxford University Press.

——, RYPKEMA, J., and BUSSER, B. (1996). The heuristic power of conceptual semantics: An investigation of spatial knowledge, in C. Tolman, F. Cherry, R. van Hezewijk, and I. Lubeck (eds.), *Problems of Theoretical Psychology*. North York: Captus University Publications, 95–102.

VARZI, A. (1996). Parts, wholes, and part-whole relations: The prospects of mereotopology. *Data and Knowledge Engineering*, 20(3), 259–86.

VIEU, L. (1991). *Sémantique des relations spatiales et inférences spatio-temporelles: Une contribution à l'étude des structures formelles de l'espace en langage naturel*. Ph.D. thesis, Université Paul Sabatier.

WAGEMANS, J. (1995). Detection of visual symmetries. *Spatial Vision*, 9, 9–32.

WHITEHEAD, A. (1929). *Process and reality*. New York: MacMillan.

WIERZBICKA, A. (1996). *Semantics: Primes and universals*. Oxford: Oxford University Press.

WILSON, H. R., and KIM, J. (1994). Perceived motion in the vector sum direction. *Vision Research*, 34, 1835–42.

WUNDERLICH, D. (1993). On German *um*: Semantic and conceptual aspects. *Linguistics*, 31, 111–33.

ZEKI, S. (1990). *Colour vision and functional specialisation in the visual cortex*. Amsterdam: Elsevier.

ZIMMER, H. D., SPEISER, H., and BAUS, J. (2001). Die Selektion dimensionaler räumlicher Präpositionen: Automatisch und nicht ressourcenadaptierend. *Kognitionswissenschaft*, 9, 114–21.

ZUKOWSKI, A., SCHWARTZ, D., and LANDAU, B. (1999). *Spatial terms in children with Williams syndrome*. Paper presented at the Boston University Conference on Language Development, November.

ZWARTS, J. (1997). Vectors as relative positions: A compositional semantics of modified PPs. *Journal of Semantics*, 14, 57–86.

——, and WINTER, Y. (2000). Vector space semantics: A model-theoretic analysis of locative prepositions. *Journal of Logic, Language and Information*, 9(2), 171–213.

Index

attention 36
Attentional Vector Sum (AVS) model 14, 111–31, 213
axial system, *see* reference (frame)
axis
  asymmetry 238
  boundary 228, 231–7, 246, 254
  dissociation with direction 26–38
  generative, major, primary or object 47, 82, 210
  prototype 228, 230–2, 237, 245, 254
  secondary, subsidiary or orienting 48
  symmetry 210, 233–5, 238
  see also direction

Boundary Vector Cell model 15, 75

coordinate system, *see* reference (frame)

direction
  betweenness of 188
  change of 86–110
  constraints in linguistic encoding of 4
  definition of 6–8, 19, 26, 70–1, 133, 135, 152, 154–5
  dissociation with axis, *see* axis
  from functional relation 261–5
  functional 161
  line 139
  linguistic devices representing 2–3, 167–8
  metric 37
  motion 168
  neurological patterns 43
  oriented 154
  part terms 52–5
  path terms 55–67
  pragmatics, role of 105–8
  projected 134
  ranges of 184–90
    definition of 189
  sameness of 171, 175
  spatial distinctions encoding 5, 12–13, 209–25
    axes 8, 18–38, 52, 111–31, 145, 170, 225
      accessibility of 140–2
      categorization of 212–13
      definition of 204
      directional marking of 220–1, 238–9, 247
      expansion along 215, 220, 225
      length of 215, 220, 223–5

  orthogonal to curved plane of symmetry 215, 220, 223–5, 247
  half-axes 227–8, 232–7, 250, 252
  half-lines 6, 43, 52, 169, 172–5, 185–90, 225
  pairs of points 6, 43, 152, 154
  regions of space 5–6
  salient object parts 223–5
  vectors 7, 12, 18–19, 37–131, 145, 169, 208, 225–6, 235–6, 253, 255
  see also reference frame
distance 7, 25, 36, 48, 50, 71, 80, 126–8, 132, 135, 138, 142–3, 145, 152, 157, 235, 255
  definition of 135, 140

map
  cognitive 69–73, 84
  Narrative/Semantic 69, 72, 84

orientation 40, 44, 49, 51–2, 61–3, 101, 113, 145, 152, 235
  change of 166–9, 177–90
  deictic 160
    definition of 163
  intrinsic or object 160, 168–70, 175–7
    definition of 161–2
  sameness of 172–5, 177

path 143, 193
  definition of 41, 44, 55, 71, 134, 138, 178–80
  shape of 63–4, 108–10
point
  definition of 134, 152–3, 171

reference (frame) 9, 11, 19–26, 29, 32, 35, 111–15, 132, 160, 170–1, 205–7, 227, 232, 248–9, 255, 258–9, 261, 263
  extraction of 237–9
  geometric constraints on 175
  absolute/environment centered 10, 21, 23, 35, 54, 100, 136, 142, 227, 230, 258, 265
  allocentric 69
  body centered 21, 22
  deictic/egocentric 10, 54, 142, 160, 227, 230, 253, 258
  eye centered (retinocentric) 21–2
  head centered 21–2
  intrinsic/object centered 9, 10, 21–3, 35, 54, 100, 160, 210, 227, 230, 233, 235–8, 253, 258
  relative 10, 54, 100, 227, 230, 253
reference system, *see* reference (frame)

route descriptions 138–9
   definition of 134, 135
route perspective
   definition of 137

shape
   change 66–7
size 40, 48–50, 59–61
space
   absolute 149
   relational 149
Spatial Feature Categorization Model 220–5, 238
   definition of 220
spatial framework theory 140–2
spatial template model 78, 111–31, 255–65
   definition of 256
   problems with 259–60
survey descriptions
   definition of 134–5

survey perspective
   definition of 137

Unique Vector Constraint 86–110

vector
   axis 47
   canonical direction 82
   direction
      definition of 101
   motion 91
   place 47
   translation 71
   vertical direction 80
   see also direction
Vector Grammar 69–85
Vector Space Model 41
visual routine 258

The manufacturer's authorised representative in the EU for product safety is
Oxford University Press España S.A. of el Parque Empresarial San Fernando de
Henares, Avenida de Castilla, 2 – 28830 Madrid (www.oup.es/en or product.
safety@oup.com). OUP España S.A. also acts as importer into Spain of products
made by the manufacturer.

www.ingramcontent.com/pod-product-compliance
Lightning Source LLC
LaVergne TN
LVHW010337260326
834688LV00036B/758